'Natural' Disasters and Everyday Lives

DIVERSE PERSPECTIVES ON CREATING A FAIRER SOCIETY

A fair society is one that is just, inclusive and embracing of all without any barriers to participation based on sex, sexual orientation, religion or belief, ethnicity, age, class, ability or any other social difference. One where there is access to healthcare and education, technology, justice, strong institutions, peace and security, social protection, decent work and housing. But how can research truly contribute to creating global equity and diversity without showcasing diverse voices that are underrepresented in academia or paying specific attention to the Global South?

Including books addressing key challenges and issues within the social sciences which are essential to creating a fairer society for all with specific reference to the Global South, *Diverse Perspectives on Creating a Fairer Society* amplifies underrepresented voices showcasing Black, Asian and minority ethnic voices, authorship from the Global South and academics who work to amplify diverse voices.

With the primary aim of showcasing authorship and voices from beyond the Global North, the series welcomes submissions from established and junior authors on cutting-edge and high-level research on key topics that feature in global news and public debate, specifically from and about the Global South in national and international contexts. Harnessing research across a range of diversities of people and place to generate previously unheard insights, the series offers a truly global perspective on the current societal debates of the 21st century bringing contemporary debate in the social sciences from diverse voices to light.

Previous Titles

Disaster, Displacement and Resilient Livelihoods: Perspectives from South Asia edited by M. Rezaul Islam

Pandemic, Politics, and a Fairer Society in Southeast Asia: A Malaysian Perspective edited by Syaza Shukri

Empowering Female Climate Change Activists in the Global South: The Path Toward Environmental Social Justice by Peggy Ann Spitzer

Gendered Perspectives of Restorative Justice, Violence and Resilience: An International Framework edited by Bev Orton

Social Sector Development and Inclusive Growth in India by Ishu Chadda

The Socially Constructed and Reproduced Youth Delinquency in Southeast Asia: Advancing Positive Youth Involvement in Sustainable Futures by Jason Hung

Youth Development in South Africa: Harnessing the Demographic Dividend edited by Botshabelo Maja and Busani Ngcaweni

Debt Crisis and Popular Social Protest in Sri Lanka: Citizenship, Development and Democracy Within Global North–South Dynamics by S. Janaka Biyanwila

Building Strong Communities: Ethical Approaches to Inclusive Development by Ifzal Ahmad and M. Rezaul Islam

Family Planning and Sustainable Development in Bangladesh: Empowering Marginalized Communities in Asian Contexts by M. Rezaul Islam

Critical Reflections on the Internationalisation of Higher Education in the Global South edited by Emnet Tadesse Woldegiorgis and Cheryl Qiumei Yu

Exploring Hope: Case Studies of Innovation, Change and Development in the Global South edited by Marcelo Sili, Andrés Kozel, Samira Mizbar, Aviram Sharma and Ana Casado

Social Constructions of Migration in Nigeria and Zimbabwe: Discourse, Rhetoric, and Identity by Kunle Musbaudeen Oparinde and Rodwell Makombe

Rural Social Infrastructure Development in India: An Inclusive Approach by M. Mahadeva

Globalization and the Transitional Cultures: An Eastern Perspective by Debanjana Nag

Forthcoming Titles

Gender and Media Representation: Perspectives from Sub-Saharan Africa edited by Margaret Jjuuko, Solveig Omland and Carol Azungi Dralega

Neoliberal Subjectivity at Work: Conduct, Contradictions, Commitments and Contestations by Muneeb Ul Lateef Banday

The Emerald Handbook of Family and Social Change in the Global South: A Gendered Perspective edited by Aylin Akpınar and Nawal H. Ammar

An Introduction to Platform Economy in India: Exploring Relationality and Embeddedness by Shriram Venkatraman, Jillet Sarah Sam and Rajorshi Ra

Unearthing the Institutionalised Social Exclusion of Black Youth in Contemporary South Africa: The Burden of Being Born Free by Khosi Kubeka

'Natural' Disasters and Everyday Lives: Floods, Climate Justice and Marginalisation in India

BY
SUDDHABRATA DEB ROY
University of Otago, New Zealand

United Kingdom – North America – Japan – India – Malaysia – China

Emerald Publishing Limited
Emerald Publishing, Floor 5, Northspring, 21-23 Wellington Street, Leeds LS1 4DL.

First edition 2024

Copyright © 2024 Suddhabrata Deb Roy.
Published under exclusive licence by Emerald Publishing Limited.

Reprints and permissions service
Contact: www.copyright.com

No part of this book may be reproduced, stored in a retrieval system, transmitted in any form or by any means electronic, mechanical, photocopying, recording or otherwise without either the prior written permission of the publisher or a licence permitting restricted copying issued in the UK by The Copyright Licensing Agency and in the USA by The Copyright Clearance Center. Any opinions expressed in the chapters are those of the authors. Whilst Emerald makes every effort to ensure the quality and accuracy of its content, Emerald makes no representation implied or otherwise, as to the chapters' suitability and application and disclaims any warranties, express or implied, to their use.

British Library Cataloguing in Publication Data
A catalogue record for this book is available from the British Library

ISBN: 978-1-83797-854-0 (Print)
ISBN: 978-1-83797-853-3 (Online)
ISBN: 978-1-83797-855-7 (Epub)

Printed and bound by CPI Group (UK) Ltd, Croydon, CR0 4YY

INVESTOR IN PEOPLE

For the 181 people who lost their lives during the 2022 Assam floods, 56 from the Barak Valley, and 45 of them being from Silchar.

Contents

List of Figures	*xi*
About the Author	*xiii*
Preface	*xv*
Acknowledgements	*xvii*
Introduction	*1*
Chapter 1 Disastrous Outcomes of Repressive Monuments	*21*
Neoliberal Weakening of the State	*26*
Misplaced Trusts and the State	*36*
Monumental Amalgamations During Disasters	*42*
Chapter 2 The Ravages of Relief Activities	*53*
Struggles within Relief Camps	*55*
Desperate Acts of the Desperate Classes	*63*
Disruptions to Life and the Sustenance of Inequality	*72*
Chapter 3 Disaster Capitalism and the Omnipresent Market	*79*
A Society of the Market	*80*
The Emergence of Petty-Disaster Capitalism	*88*
Disaster Capitalism, Corruption, and Peri-Urban India	*98*
Chapter 4 Disasters and Everyday Life on Trial	*103*
The Emergence of NGOs during Disasters	*104*
The Objectification of Lives	*111*
Capital's Routines and 'Natural' Disasters	*117*

x *Contents*

**Chapter 5   Local Politics, Right to the City, and
'Natural' Disasters**   *125*
Localism and NGOs   *127*
Communitarianism, Municipalism, and 'Natural' Disasters   *135*
Disasters and the Right to the City   *139*

Chapter 6   Towards a Humanist Politics of Climate Change   *151*

References   *165*

Index   *183*

List of Figures

Fig. 1.	The Barak River During the 2022 Floods.	4
Fig. 2.	Submerged Areas of Silchar During the 2022 Floods.	5
Fig. 3.	View of Silchar Lanes During the 2022 Floods.	6
Fig. 4.	A Tunnel on the Silchar–Guwahati Road During the 2022 Floods.	8
Fig. 5.	A Massive Traffic Jam on the Silchar–Karimganj Road During the 2022 Floods.	9
Fig. 6.	View of an Inundated House During the 2022 Floods.	18
Fig. 7.	An Inundated Road at Janiganj.	23
Fig. 8.	A Photo From Public School Road During the 2022 Floods.	28
Fig. 9.	River Water Flowing Into Kodomtola During the 2024 Floods.	31
Fig. 10.	A Child Whose Hand Had Been Fractured While Being Evacuated By His Mother During the 2024 Floods. The Mother Had to Wait Two Days for Being Able to Obtain Adequate Medicines Due to the Floods.	45
Fig. 11.	A Poster Advertising Paid Mineral Water Bottles for Being Sold at Heightened Prices on Social Media During the Initial Days of Flooding.	47
Fig. 12.	A Boat in Operation During the 2022 Floods.	51
Fig. 13.	A Relief Camp in Silchar During the 2022 Floods.	54
Fig. 14.	Inside View of a Relief Camp During the 2024 Floods.	59
Fig. 15.	Documents Destroyed By Floods in Silchar.	67
Fig. 16.	Buckets and Tubs Line Up to Get Water From a Private Tap.	72
Fig. 17.	A Helicopter Carrying Relief Materials in Operation During the 2022 Floods.	86
Fig. 18.	Dirty Water in People's Homes During the Floods.	94
Fig. 19.	A Person Walking Through a Flood-Affected House to Get Water Bottles.	95
Fig. 20.	A Belatedly Registered Relief Camp During the 2024 Floods.	100
Fig. 21.	Silchar Authority Helpline Posters Released During the 2022 Floods.	106
Fig. 22.	People Flocking to Annapurna Ghat to Check Water Levels.	134
Fig. 23.	The Flood Cell Office and the River Water Display Board at Silchar.	134

xii *List of Figures*

Fig. 24.	A Daily-Wage Worker in Action After the Floods.	146
Fig. 25.	Sofas and Mattresses Destroyed By Floods, Some of Which Were Handed Over to Dry Cleaners.	147
Fig. 26.	Post-Flood Waste in Silchar.	149

About the Author

Suddhabrata Deb Roy is currently a PhD Finalist at the University of Otago, New Zealand. He is the author of four books: *Social Media and Capitalism* (Daraja Press, 2021), *Singing to Liberation* (Daraja Press, 2023), *Pandemic Fissures* (Routledge, 2024) and *The Rise of the Information Technology Society in India* (Palgrave Macmillan, 2024). His writings have appeared in numerous journals and public forums including Capital and Class, Critique, The Sociological Review and Notes from Below, among others. This is his fifth book.

Preface

This book was conceptualised during the 2022 Silchar floods, when my family and I were entrapped within our house for more than a week with limited food, no mobile connectivity and electricity. The 12 days of uncertainty coupled with the visions of poverty that I saw during that fortnight form the soul of this book. This book should have been completed long ago. However, as I was working part-time while writing this book, the process of writing became a more gruelling one than what it already was. The most difficult part of the entire process was revisiting the difficult times that I had lived through during those fateful days and reimagining the catastrophic visions of human tragedy that the floods had laid bare which continue to haunt me even today.

I was finally able to successfully handover the final manuscript to Emerald in May 2024, specifically sometime during the middle of May. However, as a mentor once told me, it is always a difficult task to chase a moving target and so was the case with this book as well. Merely a day after Emerald sent me the manuscript queries for this book, Silchar was hit by another flood, albeit of a lower magnitude. It became necessary to include the 2024 data in this book because without that, this book would not have been able to demonstrate the argument that I am trying to make. Although this book largely talks about the 2022 floods, there are instances where this book takes recourse to narrating the incidents and stories of people affected by the 2024 floods as well.

The 2022 fieldwork for this book was one of the most challenging fieldwork assignments that I have ever had, largely because I had to conduct fieldwork in a time when there was no electricity, no mobile connectivity and a significant risk of being affected by the floods personally because my own house was under water for around 12 days. In 2024, however, the task was much easier. All in all, writing this book has been an eye-opener for me, because it allowed me to explore deeply into my own everyday reality: a reality of which I might not *directly* be a part of today but definitely continues to be something which has shaped me and affected my understanding of the society quite deeply.

Acknowledgements

A book like this which speaks directly of real-life and observed experiences can never be completed without the help of others. This book also is no exception to this general rule.

Thanks to my parents for their continued support throughout the process of writing this book.

Thanks to Marcelle for being a superb intellectual mentor and supervisor. A vote of thanks also goes to Simon and Annabel for their continued support.

Thanks to Kevin, Peter, Sandra, Kieran, Grace and Jonas for their voices of support, criticism and constant theoretical and political engagement.

Thanks to Debasreeta for keeping me motivated while writing this book. Thanks to her for helping me during the short 2024 fieldwork for this book as well.

Thanks to Sayan, Suraj, Nelson and Saraswata for their suggestions regarding this book.

Thanks to my friends, colleagues and students, conversations with whom always help me in better analysing an incident or process. The same was the case with this book as well.

Many thanks to Katy Mathers of Emerald, my commissioning editor at Emerald for this book who first told me that this was a project that they would be interested in. Thanks also to Abinaya Chinnasamy and Lauren Kammerdiener of Emerald who stood by me throughout the writing, editing and production process of this book.

At the end, thanks to all those people who shared their stories with me, without them, this book would have never seen the light of day.

Introduction

> Nodir fani katatarer uporbay jar ba Karimganj o. Ola jar jela nodi
> ee ota. Sob India Bangladesh jeno ek hoi gese![1]
> — A resident of Silchar, Assam, India, upon witnessing
> the flood waters flowing over the barbed wires that
> demarcate India and Bangladesh

Climate change and its associated 'natural' disasters are one of the defining characteristics of the contemporary world as one knows it. The United Nations International Strategy for Disaster Reduction (UN-ISDR) defines 'Disasters' as 'serious disruption[s] of the functioning of a community of a society causing widespread human, material, economic or environmental losses which exceed the ability of the affected community or society to cope using its own resources' (UN-ISDR, 2009, p. 9). The tendency of such events to occur under the contemporary developmental trajectory of free market-driven neoliberal capitalism is constantly on the rise. This is something that organisations such as the International Monetary Fund (IMF) and Oxford Committee for Famine Relief (OXFAM) agree upon (Acevedo & Novta, 2017; OXFAM, 2023). Floods, in this context, pose one of the major threats to the countries of the Global South transcending borders and other differences. As a recent United Nations Economic and Social Commission for Asia and the Pacific (UNESCAP) report states:

> The year 2022 was yet another reminder that Asia-Pacific is the world's most disaster-prone region. The major disasters of 2022 fell across the development spectrum, from floods in Afghanistan, Australia, Bangladesh, India, Pakistan and Thailand, drought in China, Kiribati and Tuvalu, typhoons Megi and Nalgae in the Philippines, heatwaves in India, Japan and Pakistan to earthquakes in Afghanistan, Fiji and Indonesia. Floods were the deadliest, accounting for 74.4 per cent of disaster events in the region and 88.4 per cent of total deaths globally. (UNESCAP, 2023, p. 1)

---

[1]Translation: 'The River water is flowing over the barbed wire in Karimganj. The water is flowing as such the entire area is the river itself. India and Bangladesh, as if, have become one!' [Translation by Author]

'Natural' Disasters and Everyday Lives:
Floods, Climate Justice and Marginalisation in India, 1–20
Copyright © 2024 by Suddhabrata Deb Roy
Published under exclusive licence by Emerald Publishing Limited
doi:10.1108/978-1-83797-853-320241001

2 'Natural' Disasters and Everyday Lives

When 'natural' disasters occur, it is the poor and the marginalised who are at the most risk of getting affected, pushing approximately 26 million into poverty every year globally (Hillier, 2018). The poor, vulnerable, and marginalised have often been at the worst receiving ends of natural and man-made disasters which have not only endangered their lives but also their sense of belongingness within the society (Deb Roy, 2024a; Iyer, 2021). In many cases, as Sainath (1996) has noted, the marginalised have been specifically put into harm's way because of the ways in which the broader socio-economic structure is put into place conforming to sociologist Ulrich Beck's (1992) proposition that capitalist development simultaneously produces wealth and risks in a manner that makes the extremely vulnerable populace – the most likely to suffer from a disaster – further vulnerable to risks.

Capitalist development is almost always uneven in nature, an exercise that is practised to favour those in possession of the resources required to utilise the effects of the developmental trajectory in a better manner (Harvey, 2005a; Wainwright, 2013). This produces, what Mezzadri (2021) notes to be, a future akin to an Orwellian nightmare. And most people in developing nations such as India *must* wake up to the grim social and individual reality caused due to this uneven development, almost every single day of their lives making the concept of uneven development is a central one for social analysts today, especially for Marxist geographers and sociologists. In India, the north-eastern region remains one of the worst examples of the uneven development which has characterised postcolonial India (Baruah, 1999; Sarma, 1966). This has, over the years, resulted in the creation of many peri-urban areas, census towns, and market towns in the region, which have remained highly underdeveloped. Urban spaces, of any kind, become the prime drivers of the problems caused by climate change under capitalism. As Mike Davis argues:

> Heating and cooling the urban built environment alone is responsible for an estimated 35 to 45 percent of current carbon emissions, while urban industries and transportation contribute another 35 to 40 percent. In a sense, city life is rapidly destroying the ecological niche – Holocene climate stability – which made its evolution into complexity possible. (Davis, 2010, p. 41)

The implications of global warming and climate change manifests itself differently in different contexts. In Assam, this manifestation is largely through the lens of the annual floods. The state has seen numerous major floods since 1947, with the most prominent ones being in 1954, 1962, 1966, 1977, 1988, and 1998 (Das & Mitra, 2003). Floods, as Das and Mitra (2003) and Das (2019) argue, cause an intense destruction within the state's overall gross state domestic product (GSDP) because of the state's increasing reliance on agricultural practices to generate income for its inhabitants. They have a devastating impact on the livelihoods of the people forcing many of them to convert to low-paid professions that reduce their well-being (Das, 2019). With the coming of a situation where environmental changes have become more rapid and unpredictable, the importance of

Introduction **3**

a predictive model has also increased manifold (Baruah, 2023a; Gu et al., 2020). However, that being said, it is also critically important to consider that merely having a predictive model would not be sufficient because the gains of technological development are unevenly distributed across the globe. The politics of climate change, global warming, and environmental degradation is immensely tilted towards favouring the developed countries, because of their growing control and domination over global finance, socio-economic, and political resources (Narain, 2017). The domination that the Global North possesses creates a situation of neo-colonialism in the Global South whereby a disproportionate responsibility is put on the latter to counter climate change and global warming (Singh, 2009). At the same time, the increasing austerity measures associated with the rise of global neoliberalism are having an immensely adverse impact on the environment – all in the name of economic progress – benefitting businesses with negligible concerns for the environment (Burns & Tobin, 2017; Sainath, 1996).

The situation in India has been so volatile that an increasing number of Indian cities situated near rivers are facing higher rates of erosion and have become increasingly vulnerable to climate threats (Kumar-Rao, 2023). Nagendra and Mundoli (2023) note that with the growth of ecologically unsustainable activities, the threat of extreme events, climatic changes, and unpredictable weather processes have risen significantly, many of which do not receive the attention that they deserve from the broader political and civil society in place. In Silchar as well, such processes are visible. The growing rates of urbanisation and the increase of commercial and residential dumping into the Barak River has caused a drastic increase in the risks of flooding in the river (Nath & Ghosh, 2022).[2] The Barak River, because of these reasons, takes a calamitous turn during the monsoons – as shown in Fig. 1 – flooding many localities, roads, and lanes of Silchar and its neighbouring areas almost on an annual basis, as portrayed by Figs. 2 and 3.

While cities of the Global North have been debating over issues concerning climate change and global warming for years, the cities of the Global South have also taken this up in recent times in different ways (Ejaz & Najam, 2023). However, as far as the highly underdeveloped regions of the Global South are concerned, the discussion about climate change has not garnered the attention that it deserves; even though as climate activist and scholar Narain (2017) writes, these regions are important parts of the struggle against climate change. This book is about one such region. This book situates itself in Silchar, one such peri-urban town situated around 343 kilometres to the south-east of Guwahati, the capital city of the north-eastern Indian state of Assam also known as the *Gateway to the North-East*. It is the prime commercial and political centre of one of the two valleys which make up the state of Assam, the Brahmaputra Valley and the Barak Valley, with the former having 580,000 square kilometres, while the latter possesses around 6,922 square kilometres (Government of Assam, 2022a, 2022b). The second-largest urban centre in the north-east, Silchar was established by

---

[2]See the report by Pollution Control Board of Assam at: https://www.pcbassam.org/RRC%20Action%20Plan%20Final/priority%20V/Barak.pdf (accessed 05.06.2024).

4   'Natural' Disasters and Everyday Lives

Fig. 1. The Barak River During the 2022 Floods. *Photo credit*: Author.

Captain Thomas Fisher of the East India Company in 1832 (Sultana, 2009) and is surrounded by rivers on three sides: Barak to its north and east, and the Ghagra to its west with the latter serving as the prime vehicle of draining the discharge flows through the natural drainage channels such as Rangirkhal, Longaikhal, Boaljurkhal, and Berakhal.[3]

The town of Silchar, which stands on the ruins of the Dimasa Kingdom of Assam, is populated mostly by Bengali-speaking people and was (and is) famously known as the Island of Peace since the days of Indira Gandhi, the third prime minister of India,[4] because of its relatively stable socio-political situation amid a more or less politically unstable north-east India. Silchar, as a town, suffers from most of the problems such as the lack of adequate drainage facilities, waste disposal systems, sanitation facilities, and water supply that most peri-urban regions of the world face (Dahiya, 2003). On normal days, these issues might seem like routine everyday matters inconsequential to the basic sustenance of human life in peri-urban regions, but during disasters such as floods, these are the issues which occupy the centre stage. The contradictions faced by people living in peri-urban regions of the underdeveloped areas become explicit during times of disasters (Davis, 1999). Floods in the

---

[3]The word 'khal' refers to a canal in Bengali.
[4]For more details, see https://www.touristlink.com/india/silchar/overview.html (accessed 05.05.2024).

Fig. 2. Submerged Areas of Silchar During the 2022 Floods.
Photo credit: Author.

region have become more disastrous considering the gradual changes which have been taking place in the region, resulting in environmental degradation which has caused a reduction in the 'tolerance, resiliency, or recoverability of the surrounding environment and reflect a concern with the assimilative capacity of the environment to absorb human intervention or to meet increasing demands' in the face of the complex social and physical transformations (Vlachos, 1995, p. 3) characteristic of capitalist developmental processes.

The period between the months of May and July in 2022 will be forever etched in the memory of the people of Silchar. During this period, Silchar witnessed the worst floods in 122 years (Tiwari, 2022), which ended up killing around 50 people

6   *'Natural' Disasters and Everyday Lives*

Fig. 3. View of Silchar Lanes During the 2022 Floods. *Photo credit*: Author.

and affecting more than a million people (District Disaster Management Authority of Cachar (DDMAC), 2022a) – *and that is only the official count*. The waves of flooding were caused by constant rainfall, which had caused the Barak River to flow at 21.59 metres, where the danger level of the river is around 19.83 metres. It is important to mention here that the highest level of the river ever recorded was in 1989, which was around 21.84 metres. Cachar receives an average annual rainfall of more than 3,000 millimetres. On 19 June 2022, the district recorded 251.20 mm – the highest in the last 10 years. Journalist Rohini Krishnamurthy (2022a) reports:

> 'The first wave of the flood was managed by the administration but in the second wave, on the morning of June 20, the district

administration was thinking about handing the rescue operations over to the Army', said Shamim Ahmed Laskar, district project officer, District Disaster Management Authority, Silchar, Cachar. The disaster levels reached L2 on June 21, requiring state intervention. L0 level is managed at the local level while the district takes over when the disaster is labelled L1. L2 would require assistance from the state government and L3 from the Centre. Flooding occurred due to the Barak river, the second-largest river in the Northeast. It originates from Nagaland and Manipur and travels for 225 kilometres along Assam before flowing into Bangladesh. The river has around 10 tributaries and five sub-tributaries. Cachar is vulnerable to flooding due to the unique geographic setting of the region, a highly potent monsoon rainfall regime, easily erodible geological formations in the upper catchments, seismic activity, accelerated rate of basin erosion, rapid channel aggradation (filling up with sediment), massive deforestation, intense land use pressure, explosive population growth especially in the flood-prone belt and ad-hoc type of temporary measures of flood control, read a 2022-2023 District Disaster Management Authority (DDMA) report. (Krishnamurthy, 2022a, para 2–6)

The two waves of flooding in Assam – the first when the river water level of the Barak was around 21.46 metres from 6 April to 12 June, and the second from 13 June to 16 September with the level of the Barak river being at 21.59 metres (Government of Assam, 2022c) – with the Barak Valley suffering the most due to the second one between 19 June until around 2 July. Most public welfare institutions continue to struggle because of the long-term effects that the floods have had on them coupled with a dwindling public-funding system, as Krishnamurthy (2022b, 2022c) states in her report. It has also initiated a string of changes within the housing patterns of individuals who have been forced to change the structure of their housing arrangements to be safe from future floods, which has become a constant fear in the region whenever there are bouts of heavy showers and thunderstorms. For example, in 2024 with merely a couple of days of rainfall during late May, there arose a massive amount of shortage in the market for essential commodities such as water bottles, candles, and vegetables with many of these commodities' prices increasing rampantly.[5] One of the other major reasons for the same is the dilapidated state of the roads and highways – including a sinking zone nearby the India–Bangladesh border – that lead to the city, which get regularly inundated during the monsoons and become unusable. And with railway tracks being washed away almost annually, trains

---

[5]For example, chilies and onions went out of the market. The prices for the available stocks rose by around 200 percent. Candles usually sold for around INR 5–10 per piece had begun to be sold for around INR 15–20 per piece.

being cancelled,[6] and flight fares ranging along the lines of INR 10,000–INR 14,000 (100 GBP – 140 GBP approximately) for a one-hour flight to Kolkata or Guwahati, the developmental issues are faced acutely by the people of Silchar during the monsoons with the valley becoming completely separated from other parts of the country, as can be seen in Figs. 4 and 5 which depict two of the most critical roads for the people and economy of Silchar.

In May 2022, the Weather Channel reported:

> For Assam, the 2022 pre-monsoon season has been much wetter than normal. Between March 1 and May 27, the state has collectively recorded 754.3 mm precipitation – a whopping 48 percent higher than its average for this period, which stands at 508.2 mm. The month of May has been even wetter, with Assam receiving 392.4 mm rain and effectively marking an excess of 56 percent compared to its May normal of 251.3 mm. (2022, para 2–3)

According to the Indian Meteorological Department (IMD), Assam had received 858.1 mm precipitation in June 2022 alone, which broke the 789.5 mm record of 1966 (Nandi, 2022). In Cachar, which was the hotspot of the floods, the rainfall for the month of June was a record 251.20 mm on 19 June 2022, in addition to it raining for 19 out of the preceeding 25 days (DDMAC, 2022a). However, intervention from the state, as the same report stated, did not begin till 22 June 2022, but by then, almost the

Fig. 4. A Tunnel on the Silchar–Guwahati Road During the 2022 Floods. *Photo credit*: Author.

---

[6]See https://www.indiatodayne.in/assam/story/floods-disrupt-rail-and-road-communication-in-assams-barak-valley-1018903-2024-06-01 (accessed 05.06.2024).

Introduction 9

Fig. 5. A Massive Traffic Jam on the Silchar–Karimganj Road During the 2022 Floods. *Photo credit*: Author.

entire region was already under water. The same was the case of May 2024, when just with around a couple of days of rainfall – mostly due to the cyclone Remal[7] – the river Barak crossed its danger level of 19.83 metres and went up to 21.54 metres – around 0.05 metres less than what it had been during the devastating 2022 floods – causing massive flooding in many neighbouring areas of the river. It is popularly believed that the town of Silchar was saved because the river water flowing from Mizoram and Manipur had largely been deviated through the neighbouring district of Karimganj because of the Longai dam breaking although there is no official or factual confirmation of the same.[8] Despite the fact that the 2024 flooding was not as severe as the 2022 floods, the occurrences of two major floods in the Barak Valley in three years point towards a massive developmental negligence that the region has been subjected to.

One cannot understand the peculiarity of the Barak Valley without knowing a bit about the state of Assam in India. Assam is a tiny state in the north-eastern part of India. It suffers from the same socio-historical, economic, and cultural marginality which is otherwise associated with other states of the north-eastern region of the country, which had often relegated the entire region when it comes to being part of important discussions. In the eyes of most from mainstream

---

[7]See https://scroll.in/article/1068675/why-cyclone-remal-left-north-east-battered (accessed 05.06.2024).
[8]See https://www.facebook.com/24BarakOutlet/videos/longai-river-dam-breaks-in-patharkandi-bazarghat-area-reportedly-followed-by-ani/486963023669819/?_rdr (accessed 05.06.2024).

10 'Natural' Disasters and Everyday Lives

India, north-east India is a region which is perennially plagued by the ills of violence and conflict (Fernandes, 2008; Sen, 2011). Assam, among all the other states of the north-east, gets relatively more widespread and diverse coverage because of its better connectivity with mainstream India. Within Assam, the region of the Barak Valley, again, suffers from further marginalisation because of the historic cultural differences that it shares with the broader Assamese society. In other words, due to the peculiar terrain that it occupies – both socially and economically – the events of the Barak Valley remain on the fringes of the imagination of the media in Assam and the north-east, which, as Raj (2016) exhibits, itself suffers from a stigmatised representation in popular media and news.

The socio-political landscape in Assam remains contingent upon the abject poverty that characterises the state. The poverty level of Assam plays a major role in the ways in which the people of the state live through, experience, and analyse floods. According to the data from the World Bank (2017), of the 31 million who call Assam home as per the 2011 Census of the Government of India, around one-third (32%) were living below the poverty line with a constant rise of consumption inequality in the urban areas (World Bank, 2017). In the Barak Valley specifically, the local news outlet Barak Bulletin's summarisation of the *NITI Aayog* report on multidimensional poverty in India states:

> In terms of multidimensional poverty, the three districts of Barak Valley have featured among the poorest of districts in the entire country. According to the report, 51 percent of the population of Hailakandi are multidimensionally poor. Cachar has 42.37 percent of the population under multidimensional poverty while Karimganj has 46.02 percent of its population under abject poverty. (Bhattacharjee, 2021, para 2)

This is not surprising considering that Assam has been having one of the lowest per capita income of the country (Chakraborty & Bhandari, 2014; Guwahati Plus, 2021). An increasing number of people in the state are dependent on agriculture for survival, and as such, most of them have come to inhabit landscapes which are very near to riverbeds, which further increases the probability of them being affected by floods.[9]

---

[9]A recent article states,

> The people of the 'floodplains' are historically used to the southwest monsoon and the floods in the low-lying areas; the minerals in the water would enrich the fertility of the land. However, in recent years, the extent of the devastation caused by floods has grown exponentially. Some reasons for this include the sedimented river water from the Himalayas combining with rain-fed water bodies in northeast India, resulting in water spilling over land in the narrow valleys, and leading to floods. The region is also prone to frequent landslides and earthquakes that deposit debris into the rivers and raise the riverbeds.

The regularity with which floods occur makes floods one of the prime socio-political questions in the region. At the same time, since the context of this particular book is in the Barak Valley – one of the two valleys in Assam along with the Brahmaputra Valley – which has been a marginalised region as far as the regional politics and its associated developmental projects are concerned. It is, however, impossible to address the question of floods in an isolated fashion. The effects that floods have on the people of Barak Valley are related to the wider everyday social questions that the region faces, both with respect to the region and with that of the nation. In the Barak Valley, floods and other natural disasters have been parts of the people's lives for generations now. An old resident of *Ashram* Road, an area mostly populated by Dalits[10] in Silchar, stated,

> Floods have been a part of my childhood, my youth and now, even my old age. Earlier when floods used to occur, we would rush to the high lands with everything that we have with us. Food, clothes, everything. Today however, things are different. You see those high lands are now covered by buildings. Rich people live there today.

Disasters such as earthquakes, floods, cyclones, and the like not only leave the people with economic losses but also cause an irreparable damage to the psychology of the affected. They affect certain everlasting impressions on the ways in which the interaction between the individuals and the society structures itself.

The havoc that floods bring in within the region gets further aggravated by the rising poverty and its associated helplessness. This helplessness, in recent times, has been accentuated by the rising precarity that these people face due to their linguistic distinctiveness with the broader Assamese community in the state. The fear that such processes have instilled in the minds of most of the people of the valley had an effect on the way in which they perceived the floods. As a street vendor selling goods such as oil and candles on a tiny plank by the road stated:

> That day, my shop got washed away. I used to give a tiny tea shop right here, but now I am forced to give this shop on this tiny plank. All the tea powder, milk, sugar, and the equipment have been lost. I do not know how I will survive through this calamity.

These heavily sedimented rivers cause soil erosion, resulting in further flooding. Compounding these issues are several man-made factors, such as construction on high seismic zones, rampant deforestation, cutting down of hills, clearing of forest land for agriculture, unplanned urbanisation, destruction of biodiverse wetlands, and the pressures of a booming population on a fragile ecology that have made the floods much worse than before. (Barthakur, 2022, para 6–9)

[10]Dalits here refer to the Kaibartas and other scheduled caste communities. In general, colloquial language, however, they are not referred to as Dalits by people of Silchar, but rather the term 'Kaibarta' is specifically used.

12 *'Natural' Disasters and Everyday Lives*

The street vendor, interestingly, was also a Muslim and had been particularly affected by the narrative of *Flood Jihad* which had been making rounds in the town then.[11] Natural disasters in the region often become secondary issues when compared with the other major issues occurring in the region. For example, recently, the state has been in the news for many reasons apart from 'natural' disasters such as floods. The debates surrounding immigration and citizenship in recent years have fuelled the regeneration of tendencies such as ethno-nationalism and sub-nationalism in the region (Baruah, 1999; Mander & Singh, 2021). The point behind putting this here is to argue that fear works in a variety of ways among people, especially if they come from the marginalised section of any society (Davis, 1998). Fear, as Tudor argues, is both emotional and cultural in nature, even though:

> Traditionally fear has been understood as one of the emotions (often, indeed as a 'primary' or 'basic' emotion) and thus consigned to the tender mercies of psychology. Contrary to sociologists' worst stereotypes, however, that has not meant that the prevailing conceptualisation of fear, and of emotions more generally, has been little more than a behaviourist gloss of an emotional 'black box'. (Tudor, 2003, p. 241)

Staupe-Delgado (2022) and Baruah (2023) have exhibited, in detail, the dehumanising and destabilising effects that an anticipation of a disaster has on the everyday lives of a community. When the fear of an impending disaster acts in conjunction with existing socio-political fault lines, it produces, what Craib (2010, p. 291) mentions to be, a situation where 'personal history and personality structure[s come to have] the greatest force, combing and directing the other factors' of fear. Natural disasters and their associated impacts are related to the wider socio-political events experienced by the people and the space they inhabit (Davis, 1990). The effects that a particular disaster has on a community are related to the processes of social, political, and cultural transformation that a community is going through at any given point of time. The people's movements in the state regarding the debate about citizenship are important ones from the perspective of the social sciences, but the horrors of the floods, on the other hand, penetrate deeper because they happen almost on an annual basis. Moreover, there is an intrinsic connection between the two which needs to be dialectically analysed, something that is explored all throughout the present book. Most of the people affected by floods in the region are almost certain that it is something that will be repeated, and as such they need to prepare themselves for the same. It is the fatalistic belief in their own troublesome fates that is captured in this particular book and analysed sociologically using a Marxist framework.

Aijaz (2019) argues that the problems faced by peri-urban regions need intervention at a policy-level to overcome the problems which plague them. This

---

[11]'Flood Jihad' is a term that was used to refer to the idea that Muslims are responsible for the floods of Silchar in 2022. For more details, refer to https://www.bbc.com/news/world-asia-india-62378520 (accessed 06.05.2024).

Introduction   **13**

includes the problems associated with the methods through which the government classifies the areas which often overlook the existent local peculiarities. The constant marginalisation of these spaces results in them suffering from issues which result in having low levels of service assurance and delivery mechanisms (Mondal, 2021). This problem becomes acute in places such as the Barak Valley, which are at the receiving end of a bourgeoning socio-political crisis. When disasters occur in these spaces, they get affected by a combination of adverse conditions such as governmental mismanagement, lack of coherent mechanisms to manage extreme climate events, and inadequate land-use policies (Maheu, 2012). All of these get accelerated when the region in question also simultaneously suffers from political marginalisation, like the Barak Valley in Assam, so much so that Sharma and Gayan's (2014) recent study of flood mitigation in Assam does not even have a single mention of the 'Barak' or the 'Barak Valley'.

Many regions in north-eastern India repeatedly get inundated by river water flowing into the landed territories every year, especially Assam. It is common that floods and other such natural disasters, whenever they happen, are widely reported in the media. The situation becomes grimmer when one considers the floods occurring in Barak Valley, a region which has been historically and socially marginalised within the state because of the linguistic and cultural differences that the region possesses in comparison with the broader social structure and characteristic of the state of Assam. This distinctiveness of the region has played a key role in the uneven development that the region faces, both economically and socially (Prasad, 1988). Disasters have different dynamics in different spaces. Some places are structurally ill-suited to withstand harsh effects of disasters than others (IMF, 2016). The way in which a particular region is placed, economically, socially, and politically, affects the media coverage of the region – establishing the validity of Marx's (1867/1976) arguments surrounding the all-encompassing nature of capital. Be it Chennai or Mumbai, floods have enjoyed an enormous portion of the media space in the national mainstream media which includes widespread prime-time news coverage and even movies and the literary non-fiction space (Ge, 2019; Viju, 2019). However, when it comes to states like Assam and Manipur, which exist at the fringes of mainstream India, both geographically and politically, the reportage also becomes highly marginal in nature. Even with the devastations that the recent flood had caused resulting in affecting around 3 million people and killing around 12 *officially* (India News, 2022), the *Times of India* – the most widely circulated newspaper in India – did not carry a single headline.

The '2022 Silchar Floods' remained an obscure event for most of the widely circulated news agencies of the state. On conditions of anonymity, a journalist with a local media outlet told the author:

> The events of this region never get reported properly in the state newspapers. Check the coverage in newspapers such as *The Assam Tribune* or *The Sentinel*, and others. It is always written as something that is merely an accident, while in reality the problem was, is and will be much severe.

14 'Natural' Disasters and Everyday Lives

The point that the journalist was trying to raise is that floods in the Barak Valley are a structural issue and as such demand interference from higher authorities at a much more structural level. Marx here becomes a useful philosopher to consider because he based his theories on the fundamental aspect of social change. Using Marxist ideas enables one to theorise the perils which capitalist development poses to the human society in terms of the segregation that it produces even within natural artefacts, in addition to producing, as Foster (1999, 2000) has argued, a metabolic rift between nature and human beings.

The extremely low coverage of the floods caused by Barak also speaks volumes about the classification that even natural bodies such as rivers pass through under capitalism. Reputed scholars have written at length about the Ganga (Mallet, 2017; Sen, 2019), and even the Brahmaputra (Choudhury, 2021; Saikia, 2019), but the same has not been the case with Barak or even the Yamuna. The issues surrounding the Barak and the people who live through or by the river have often been relegated to *hard-to-find* government files. One of the few times when the river comes to the attention of the mainstream national news is when cases surrounding something important to the global political or scientific community comes up. One such example can be cited of the Ganges River Dolphin, which is facing the risk of extinction in the Barak Ecosystem (Ghosh, 2019), which has received widespread attention in the national media. The attention that the Ganga gets is easily justifiable considering that it is one of the very few major rivers – *perhaps the only one* – flowing through the country which has its origins *within* the country (Nehru, 1946/2015).

The fetish with the Ganga – both for its geopolitical and religious contexts – raises important questions about the relationship that human beings share with natural entities such as rivers. If one looks at the situation rationally, then the river Barak is no less important than the Ganga, the Yamuna, the Brahmaputra, or the Godavari – all of which have featured in Jawaharlal Nehru's (1946/2015) seminal study on the dynamic relationship between rivers and the Indian civilisation, *The Discovery of India.* Saikia's (2019) work on the *Brahmaputra* too focuses on the intrinsic relationship that the river shares with Assamese civilisational values. Choudhury's (2021) work brings out the conditions created by floods due to the Brahmaputra in the state, which affects almost the entire state.

Sharma and Gyan's (2014) research detail out the gradual widening of the Brahmaputra River from 3,870 square kilometres during 1912–1928 to around 6,080 square kilometres in 2006. That being said, even the Barak is no less significant in terms of numbers and the diversity of the regions which it affects:

> The Barak sub-basin drains areas in India, Bangladesh and Burma. The drainage area … lying in India is 41723 sq.km which is nearly 1.38 percent of the total geographical area of the country. It is [surrounded] on the north by the Barail range separating it from the Brahmaputra sub-basin, on the east by the Na Lushai hills and on the south and west by Bangladesh. The sub-basin lies in the States of Meghalaya. Manipur, Mizoram, Assam, Tripura and Nagaland. (Government of Assam, 2022a, Para 3)

Barak has not received the attention it deserves considering that it affects not only regions within India but also flows into Bangladesh by bifurcating itself into the *Kushiyara* and the *Surma*, thus becoming one of the few rivers in the country the management of which directly affects India's foreign relations with its neighbours. There is no extraordinary academic or non-fiction book on the river Barak or the people who live through and beside it, which can be partially attributed to the marginalisation that the region and its people has been suffering through since independence. Hence, it is unsurprising that the floods in the river Barak have also remained elusive to most academic and social commentators.

The 2022 floods in Barak Valley were different from other years, for two reasons. The first reason being that this was one of the rare occasions in which there were multiple floods in quick succession. The second, and the more critical one, being the fact that numerous dams around the town of Silchar and the neighbouring town of Karimganj – a quaint little town placed around the India–Bangladesh border – broke down due to the heavy rainfall in the region accompanied by the rising water level in the Barak and *Kushiyara* – a distributary of the Barak which flows into Bangladesh – rivers.[12] During the first wave of floods in 2022, the losses incurred by the people in the urban region were mostly nominal because the water merely scratched the surface of the urban area:

> We were not much worried when the water came in the first time. Most of us thought this was only the regular floods occurring in the region. But, when the water came in for the second time, it came fast, and it destroyed everything that came in its way.

The same kind of narrative can be found in many pieces which have appeared in digital news outlets, especially those coming from citizen journalists.[13] Digital news media has been an important aspect of the ways in which the news of the devastating floods have reached the broader masses of the nation. In contemporary societies, the relationship between technology and human societies is something that has captured the attention of social scientists for decades now. Technology has come to be one of the major aspects of life under capitalism as one knows it, especially in this day and age of social media (Deb Roy, 2021). Indeed, it was through technology that when the water entered the towns of Silchar and Karimganj for the second time since May 2022 in June, the news spread rapidly across the town. The first news about water entering through the embankments came through videos produced by local people living near the dams on their mobile phones. The videos showing the river water entering the human-inhabited regions at a pace akin to water flowing down a fully active monsoon

---

[12]See https://waterresources.assam.gov.in/portlet-innerpage/barak-river-system (accessed 05.05.2024).

[13]For example, see https://www.thequint.com/my-report/assam-floods-silchar-inundated-in-flood-water-no-electricity-food-shortage#read-more and https://www.outlookindia.com/national/assam-floods-2022-diary-from-silchar-news-206808 (accessed 05.05.2024).

16 'Natural' Disasters and Everyday Lives

waterfall caused terror among many of the townsfolk. The flood water entered the town of Silchar on 18 June and Karimganj on 19 June for the second time in merely 26 days (Zaman, 2022). It must be mentioned here that exact dates are difficult to be cited in such cases because the flow of water had been haphazard and, at times, quite irregular. A resident of College Road, a colony dominated by the neo-middle service class, stated,

> The floods were horrible this year. In my lifetime of more than 40 years, I have not seen a flood as being as this one. The early 90s' floods were previously the worst, but even at that time, the water had not flooded places that it did during this time.

By June, the town had already been reeling under the effects of the flood which had happened in the outskirts of the prime urban landscape – the hallowed space for around 3 million people as per the census data of 2011.[14] During the second wave of the 2022 floods, the rainfall went about incessantly for close a period which exceeded 10 days, adding to the woes of the residents of the area. The water had reached such a height that boats became the only means of transport for treading distances as short as 200 metres. A teen-aged girl recounted:

> It is becoming increasingly difficult for us, especially women. The roads are not safe. These boatmen, I do not mean to be rude, but I do not know them. Also, the prices are too expensive. I had to pay around 200 rupees for what might be only 250 metres.

In the towns of Barak Valley, boats entering the town is a common phenomenon – almost an annual one. The ones who drive the boats are usually local people from the outskirts of the urban areas, who live near the river through meagre incomes generated through daily wages or by ferrying people and goods across the Barak.

However, during floods, the same people are found to be engaged in multiple kinds of jobs, which often includes selling commodities at higher prices to urban residents much like the local shopkeepers do in times of distress. The question that looms large at this point is whether these poor boatmen are mere disaster capitalists (Klein, 2007) or vengeful *class warriors* who desire to extract revenge on the urban people who have neglected them for decades considering them to be sub-human beings. That is the prime question explored in the third chapter of this book. Floods have a wider socio-economic and cultural impact on the communities which they affect. On the economic front, floods raise questions about public services and the ways in which reforms impact individuals and the community as

---

[14]The Barak Valley is constituted by three districts, namely Cachar, Karimganj, and Hailakandi. See https://cachar.gov.in/information-services/district-at-glance, https://www.census2011.co.in/census/district/158-karimganj.html, and https://www.census2011.co.in/census/district/153-hailakandi.html (accessed 05.05.2024).

a whole. On the political front, they raise certain questions about the character of governance and polity in the region, especially when floods have been on the manifesto of numerous political formations – including the current ruling government – in the state. On a more philosophical level, floods point towards the very core of our human nature – the Darwinian struggle between co-operation and competition, something that Marx (1844/1973) had also paid great attention to in his explorations towards conceptualising *the new human.*

Environmentally, regular floods make it imperative for us to think about the ecological footprint that human settlements have on the environment. Sociologist Ted Benton (2018) notes about the importance that Marx still holds at this critical juncture of human progress especially through the relationship conceptualised by him between feudal and capitalist societies – with the former being exploitative and deeply unequal yet lacking the destructive expansionism of industrial capitalism. These conditions ensure that contemporary urban spaces, especially the peri-urban areas, remain under a constant pressure for land resources, resource utilisation, migration, etc., perhaps more so than mainstream urban centres because of them being not fully rural or not fully urban and thus getting overlooked (Tiwari, 2019). 'Natural' disasters, as this book will argue in detail, provide a unique argument in favour of certain aspects of life that neoliberal capitalism does not look kindly upon. For example, urbanisation processes in Barak Valley have not looked kindly upon the hillocks, commonly known as *Tilas* in the region, which populate a significant portion of the valley. These *tilas* during the floods, act as saviours of thousands of people only to be forgotten as the water recedes. The areas usually serve as temporary refuges for a lot of people. However, of late, most of these hillocks are in the process of getting destroyed for the sake of building up apartments and for serving up as land-fillings for low-lying areas.

With the rendering mainstream of gentrification, the urban space under contemporary capitalism, has come to be characterised by class contradictions, which is increasingly causing a rise in the magnitude of the adverse effects that disasters have on the marginalised sections of the population, anywhere in the world (Diwakar, 2021; Stein, 2019). The problems that people face due to the floods vary with respect to the social positioning that one possesses. For example, when floods happen in India, or for that matter in most places across the globe, the first service which gets disconnected *almost immediately* is electricity. For most of the people living in flood-inundated houses – one of which is shown in Fig. 6 – electricity is a risk, but they still continue to desire for electricity because of the way in which electricity has become the very core of human activity in contemporary society. Natural disasters provide an interesting avenue for analysing the interactive relationship that technological developments share with the human society.

This book mainly draws from the analysis of around 100 in-person interviews conducted mainly during the period when the town of Silchar was inundated by the 2022 floods. These interviews were conducted with people affected by the flood, officials and volunteers associated with relief work during the floods, as well as with community leaders – all of whom have approached the issue with a unique lens of their own. The interviews which have been used in this book were

18 *'Natural' Disasters and Everyday Lives*

Fig. 6. View of an Inundated House During the 2022 Floods. *Photo credit*: Author.

mainly conducted in three languages, Bengali, Assamese, and Hindi. However, in this book, these interviews have been translated to English for the ease of the reader. This book is divided into five chapters, along with a short conclusionary chapter. This introductory chapter has briefly traced the dynamics of floods in the state of Assam, and especially within the Barak Valley, thus laying the foundation of the succeeding chapters by putting the socio-economic and political context in which the work locates itself.

The first chapter of this book discusses the role of the state and its continuing relevance for the people of underdeveloped regions. This chapter discusses the neoliberal assertion on the Indian state and the implications of the same during times of natural disasters in highly underdeveloped regions. This chapter problematises the idea of the state and its relationship with communities and the disaster recovery frameworks. It focuses on the dialectical relationship that individuals share with society and the state as had been highlighted by Marx (1973) and how it affects the marginalised people during times of distress such as floods and other natural disasters. The second chapter discusses the dynamics of the relief camps. This chapter takes the readers to the relief camps which are set up by the state to help those affected by the floods. These relief camps are usually set up in various schools, colleges, and other educational institutions. Drawing from interviews conducted within relief camps, this chapter brings out the ways in which an alternate economy is often set up in these camps. The second chapter, in doing so, engages with the processes affecting the creation of a petty-disaster economy in the region where a group of 'small entrepreneurs' often labelled as 'black marketeers' – who are otherwise workers – rise up to extract profits out of a disaster. For example, during the recent floods in the state, there occurred a widespread crisis of essential goods, including drinking water as many narratives in this book exhibit, which was used by many flood-stricken people to commercialise the same – some out of need and some out of greed. The fourth chapter seeks to unearth the dialectics between these two drivers of human action during times of disasters. The fifth and final chapter engages with the ways in which people tend to rediscover their sense of being human within extraordinary circumstances such as floods and earthquakes. This chapter focuses on the activities that people engage with to spend time during the floods, which include fishing, initiating community gatherings while sitting on balconies and elevated platforms, etc. It highlights the absence of certain so-called omnipresent aspects of modern living such as technology, commodities, and services and how it affects community life in peri-urban regions.

Analysing floods in the Barak Valley needs to take into account the fact that the fear of an imminent disaster gets amplified for the people from the marginalised section of the society because of their already existing fears and anxieties which shape a major part of their everyday lives (Furedi, 1997/2002). These fears share a dialectical relationship with the social contradictions which are constructed by both macro and micro socio-political developments. Tenhunen and Saavala (2012) have noted that one of the key contradictions faced by governance in India is mediating the relationship between the centre and the states because of its historically mixed system of governance, which have become further complicated after the coming of the neoliberal reforms in 1991. The situation becomes extremely complicated in places such as the Barak Valley, which remain on the fringes of both the central and the state governments – a place which neither 'British democratic parliamentarism [nor] Soviet-styled planned economy, which strengthened the power of the Central Government' (Tenhunen & Saavala, 2012, p. 53) could put on the path of rapid development. Poverty still remains one of the most glaring problems of the valley as has exhibited by Ghosh (2013) with many households still suffering from lack of access to safe drinking water, proper electricity facilities, etc., which drastically affects the human development in the valley. Low human development, in turn, affects the health of those coming from the marginalised sections of the population, which is further put at risk by floods as Phalkey et al. (2010) shows. This is not only the case of India, but in the entire Global South itself, floods and other natural disasters share an intrinsic relationship with poverty and the health of the poor (Dube et al., 2018).

20 *'Natural' Disasters and Everyday Lives*

The history of flood-related control measures in Assam is an old one as Kumar (2021) and D'Souza (2016) have shown in their recent research. But both have also argued that these measures had been established taking into context a framework which was extremely colonial and undialectical in nature such as the overwhelming importance placed on embankments. Natural disasters and the response towards them cannot be seen as completely distinct from the socio-political and economic situation in which a particular community or an individual finds itself or oneself in. Marx (1973) had argued that one of the defining characteristics of capitalism was the way in which public services are separated from the state and are then subsequently left to the market. This becomes further aggravated in situations such as the present, when with the increasing magnitude of the effects of climate change, floods and other such natural disasters are only determined to increase in the coming days (Jha et al., 2012). The fear of such disasters not only affects people living under their threat at an individual level but also at a social and community level, often directing the kind of relationships that they share with the state, society, and community. This book will show that despite the apparent failure of many 'Marxist' states to address the degradation of the environment (Cole, 1993), a Marxist approach still holds a lot of potential in addressing the climate catastrophe that humanity faces. But to do that, one needs to unearth the dialectical Marx from the crude one who has come to dominate contemporary social theory. This book is an attempt in that direction.

Chapter 1

Disastrous Outcomes of Repressive Monuments

'Natural' disasters occur at regular intervals, and have become more frequent under contemporary forms of extractive capitalism. Some people take them to be divine signs regarding the future of humanity on planet Earth, while some others believe them to be an indication of the degradation that rampant human activities have caused on the planet. Both these versions of explanations become greatly relevant in the minds of people, especially during an apocalyptic event such as Covid-19. The floods of 2022 were nothing of the scale of Covid-19, but they still had a lot of similarities. The first one among all the similarities being that both these events turned into human catastrophes for a majority of the people who came into contact with them. Both of them – the first one at a grand scale and the second one at a smaller *yet highly impactful* scale – turned into human catastrophes because of the inefficient nature of the governments involved in mitigating the effects of the same. The conversion of a natural disaster to a human catastrophe depends intimately on the nature of governance present in any society. As Kelman argues in a recent book:

> Disasters are not natural. We – humanity and society – create them and we can choose to prevent them .... Stating that natural disasters do not exist because humans cause disasters seems insanely provocative. We witness nature ravaging our lives all the time: from a city underwater after a hurricane to rows of smouldering houses after a wildfire to the dust rising from the ruins after an earthquake. (Kelman, 2020, p. vii)

A catastrophe is a combination of human actions within the Anthropocene[1] and the political policy-making decisions that neoliberal bourgeois politicians make in the wake of the effects that any natural disaster brings forward (Wallace

---

[1]The term denotes the current geological age that is often viewed as the period during which human activity dominates over climate and the environment.

'Natural' Disasters and Everyday Lives:
Floods, Climate Justice and Marginalisation in India, 21–52
Copyright © 2024 by Suddhabrata Deb Roy
Published under exclusive licence by Emerald Publishing Limited
doi:10.1108/978-1-83797-853-320241002

22   'Natural' Disasters and Everyday Lives

& Wallace, 2016). To find solutions amid such an adversely manipulated socio-political and ecological situation, it is necessary that one moves beyond solutions proposing merely superficial alterations and moves towards systemic changes. A natural disaster, especially its causes and effects, demands a materialist analysis soundly based on scientific and social scientific analysis because such analysis is based on sound data and analysis critical to a holistic study of the disasters (Phukan, 2003). Natural disasters, however, have frequently been believed to be markers of the wreath of the Gods. When the floods hit the Barak Valley in 2022, these kinds of explanations were heard quite in plenty. A teenage boy said sitting on the back of a *thela*[2] was seen gossiping to his friends: 'You know, the major issue is all this silting that we are doing at the Barak river. The river has a capacity. You exceed that and the river fights back. That is why we are seeing all this devastation going on'. From a quite different social position, an older individual – a Muslim man – roaming the streets during the floods, kept on murmuring about how the floods signal the wrath of the Gods:

> The floods are a symbol of destruction. The world as we know is coming to an end. This world is rotten, its coming to an end. I am good, God will save me, I will escape. All these roads will go, all these houses will be washed away. This is my prediction, *O God*, save me. These humans have ruined the good Earth that you had created.

These words from a wandering homeless person came in the wake of a situation when the Barak River was coming down at a gushing speed onto its closest inhabitants, the poor working-class settlements in places such as Rongpur and Dudhpatil.[3] The two waves of flood that hit Cachar during the 2022 summers took a certain time to reach the shores of the privileged. This is because most of the relatively well-off do not occupy spaces that are closer to the river or in low-lying areas such as parts of Janiganj, Malugram, Rongpur, Link Road (Second), and Sonai Road. The 2024 floods again prove this, because these were the very same areas that again got flooded during the initial days of the floods in May 2024, with displaced people from Kalibari Chor in Janiganj – the road leading to which is shown in Fig. 7 – and other low-lying areas in Malugram having to rush to relief camps.

Those higher up in the hierarchy believe natural disasters to be avenues through which they can make their presence and importance felt in the society. This is done mostly through the economic impacts that their presence or absence has on the society, especially among the highly marginalised communities. For example, despite industries shutting down *mostly* during any major disaster, their presence and importance in the society continue to be felt with rising unemployment and loss of income-generating activities for the people during a major disaster. This

---

[2]A kind of hand-operated pull cart usually found in India.
[3]Areas situated by the riverbank in Silchar.

Fig. 7. An Inundated Road at Janiganj. *Photo credit*: Author.

is because the access to resources in different parts of the world acts as a major factor in determining the extent to which any populace can resist the effects that natural disasters have on them. A properly functioning publicly owned welfare mechanism helps the highly vulnerable and marginalised populace to gain access to certain services more efficiently during disasters because it is often the case that such services do not only provide people with commodities and services but also with a sense of empowerment capable of transforming them into more assertive and participant citizens.

The process through which most of the residents of Silchar analysed the floods was influenced by a mix of developmental negligence (that the Assam government has been exhibiting towards the Barak Valley)[4] and electoral apathy. The Barak Valley does not feature prominently within any national or regional discourse and

---

[4]The developmental negligence towards the Barak Valley stems from decades of socio-political strife between the Assamese and the Bengalis, which has also affected the Bengalis settled in other parts of the north-east such as Shillong and Itanagar (Choudhury, 2023).

24   'Natural' Disasters and Everyday Lives

suffers from a constant lack of focus by the authorities on the problems that the area faces. While in 2022, the administration and authorities stated that the floods caught them off-guard at a sudden moment – which is also difficult to believe because of the highly improved weather prediction mechanisms in place – they did not have a better reason to provide to the people during the 2024 floods as well with the same Bethukandi and Shibbari dams being the key causes behind the floods.[5]

The lack of governmental attention towards this critical dam points towards a persistent negligence of the state authorities. The constant negligence that the Barak Valley has faced from the state government of Assam because of it being a region dominated by Bengali-speaking people has led to its relative developmental stagnancy, so much so that even in recent historical accounts, the name of the Barak Valley occurs only a few times, and the region is reduced to merely a cursory reference. Assam, as a state, has never looked very kindly upon the Bengali population that lives within its southernmost region. There has been a long history of structural and material violence that has been at the forefront of dictating the kind of relationship that the Barak Valley shares with the broader Assamese state. The persecution that the Bengalis faced in Assam has been such that, as Deb states:

> [The Bengalis] came to be identified and perceived as threats by the Assamese fearing they would disrupt the cultural dominance of Assam and accordingly, the post-independence period, which saw the Assamese in Assam rising to power, using it to ensure that the Bengalis in the region are not given a chance to assert their supremacy and dominance and the decision to implement Assamese as the sole official language in 1960 and the medium of instruction in 1972 in order to ensure that the sons of the soil are benefitted, can be considered as the ways in which such an end was sought to be achieved by asserting their Assamese cultural identity. Hence, the power position of the Assamese was used to consolidate their position in the Assamese society and culture. Also, violence against the non-Assamese speaking population continued from time to time creating an environment of panic and terror for them, quite similar to the kind of environment that prevailed under Hitler in Germany. (Deb, 2019, p. 45)

The popular imagination of Assam, following such a hostile environment that continues to prevail between the Bengalis and Assamese, remains restricted to the Brahmaputra river and the spaces associated with it. The Barak Valley – that

---

[5]See https://www.barakbulletin.com/en_US/breaking-bethukandi-sluice-gate-breached-river-water-flowing-into-mahisha-beel-men-at-work-will-control-water-resources/ and https://theshillongtimes.com/2024/05/31/assam-minister-takes-stock-of-barak-valley-flood-scene/ (accessed 02.06.2024).

constitutes 8.2 percent of the entire population of Assam – does not feature in within the mainstream Assamese or north-eastern discourse so much so that in Samrat Choudhury's (2023) acclaimed history of north-east India – an extraordinary work dealing with a suppressed region – the word 'Barak' only appears six times and mostly in contexts which do not problematise it as being a part of north-east India but rather as an anomaly. This is true for every important aspect of the modern socio-political fabric of the state, including the treatment meted out to 'natural' disasters in the Barak Valley. The analysis of floods in the Barak Valley performs the critical role of unearthing the local dynamics of a neglected yet critically important region, which has hitherto been stripped off the benefits of a major socio-political analysis. Floods are an important source of such analysis because both the Brahmaputra valley and the Barak Valley have them in common. Both these valleys get regularly inundated by floods. The negligence is criminal because out of the 181 people who died due to floods in 2022 in Assam, 56 of them were from the Barak Valley – 45 of them being from Silchar (Government of Assam, 2022c).

Floods are one of the most common natural disasters globally. However, academic scholarship usually remains restricted to sites of prominence. Even among countries of the Global South, the scholarship on floods remains scarce and rarer in the context of the highly underdeveloped regions of the Global South. Natural disasters affect the everyday lives of the people, especially those coming from marginalised social and geographic backgrounds. An emphasis on the everyday lives of these people and their relationship with the state engaging with questions of housing, community, belongingness, social fears, anxiety, and the like exposes the most innate and subconscious emotions which are invoked among people when they have 'water in their homes' and how it is related to the broader relationship shared between normalised human existence under capitalism in the lives of people from the *South of the Global South* (SGS). The SGS in this context refers to the non-mainstream part of the Global South. It refers to spaces such as the Barak Valley which despite being a part of the Global South do not even receive the minimum requisites of human and economic development within the Global South and become, in the truest sense, the SGS.

The activities that people engage in the SGS are often lost within the myriads of histories written from the perspective of the developed regions, even among those of the Global South. They do not adequately reflect the grave consequences that climate change and global warming have on the underdeveloped regions and the people therein. Regions such as the Barak Valley and cities such as Silchar reflect the critical implications of the blame-game surrounding global climate change – with the developed nations blaming the underdeveloped and developed ones and vice versa (Narain, 2017) – on the most vulnerable populace of the world. Natural disasters not only have an economic but also a psychological impact on the people, many of which prove to be instrumental for newer conceptions of institutions such as the state, the community, and social relationships to come up, which makes it important that one not only engages with the social impacts of disasters but also with the psychological effects that they have on the people. The state in the context of India continues to occupy an important position in the

26 'Natural' Disasters and Everyday Lives

response to any disaster, because of the control that it continues to exert in the Indian socio-political scenario within the domain of legality and provisioning, despite the neoliberal assertion. This chapter problematises the idea of the state and its relationship with communities and the disaster recovery frameworks. This chapter begins from a discussion on the neoliberalism's effect on the state in India and follows up with a section on the dependency that people from the SGS continue to have on authoritarian structures such as the state. The last chapter, using Lefebvre's theoretical position on monuments, argues that in the SGS, the community does not provide an opportunity for radical restructuring but is rather a part of the statist framework.

Neoliberal Weakening of the State

Human society has now become prone to disasters with the growth of apocalyptic tendencies, especially for people from lower socio-economic backgrounds. Most of the marginalised today live a life that remains bereft of much protection from the effects of climate change and, as Ulrich Beck (1992) had highlighted, live in the constant fear of getting exposed to risks. For many of these sections of the populace, natural disasters have become a part of the annual cycle of life and sustenance because of the increased risks that they have been exposed to because of the growing repercussions of climate change and global warming. Climate change, as Holthaus (2020) argues, has compounded the effects of natural disasters giving people lesser and lesser time to regroup and plan their response to the same. They have no longer remained a matter of luck, or ill-fortune, but have rather become a part of the capitalist framework itself creating more vulnerability for the already vulnerable. The highly marginalised remain more at the risk of being harmed by ecological issues because they lack the necessary social capabilities that are needed to counter these effects. The risk society that has been brought forth by contemporary capitalism has produced a complicated sense of modernity for most of the people, as Beck writes:

> In advanced modernity the social production of wealth is systematically accompanied by the social production of risks. Accordingly, the problems· and conflicts relating to distribution in a society of scarcity overlap with the problems and conflicts that arise from the production, definition and distribution of techno-scientifically produced risks. (Beck, 1992, p. 19)

One would have ideally assumed that with the development of science and technology, the world would have gradually moved beyond these tendencies avoiding them to the best of its capabilities. However, as the developmental trajectory of capitalist development has exhibited, the world has gradually moved – instead of moving ahead of the impending apocalyptic tendencies – towards their possible realisation at an increased pace. Natural disasters characterised by extreme climatic events caused by climate change are a grim proof of this transformation of the human society. The effects of climate change have been such that even upscale neighbourhoods in many Indian cities have now become vulnerable to floods and other natural disasters. For example, upscale neighbourhoods in Silchar such as Ambicapatty in Silchar have become vulnerable to situations becoming akin to floods with a few hours of rainfall. A resident from one of the posh neighbourhoods reminiscing about the floods of 2022 informs:

> In Ambicapatty, usually floods come very late if it ever happens. But this time it was different. Most of the houses have done very deep drillings. The other is the issue with apartments and societies with very deep foundations which have cropped up everywhere. They have depleted the soil quality and the height of the road, so the water today comes in easily to our homes. While previously, we used to have water after a round of heavy showers, but that was only on the roads, today with 2-4 hours of rains, it enters our houses on the ground floor though a little.

Similar statements could be found in other upscale neighbourhoods such as the Public School Road and Hospital Road–Central Road stretch. The conditions of the former, i.e. the Public School Road, are shown in Fig. 8. These neighbourhoods and areas were previously relatively unaffected by floods – at least those parts where middle-class people lived – as another resident from such a neighbourhood argued, 'Earlier, we never had to worry much. Yes, floods would come and go and my house would also be affected. But the water would just about enter and leave. But nowadays, if it enters, then it stays for long. Sometimes, it even stays for more than a week'. These changes point one towards the changes in the nature of the effects that floods have. In Assam, floods are a regular feature of the everyday reality. Being an agriculture-based region,[6] the impact of floods in the region has had tremendous effects on its economy and culture.

Disasters alter the nature of the social relationships that one finds oneself engaged in, often making them more individualistic in nature. They reinvigorate the relationship between individuals and the state, in such a way that people feel an inert need of the state because they see the state as being an all-encompassing structure that is supposed to work for their welfare. During the pre-neoliberal years, the state had come to be the most important facet of the lives that individuals lead in most of the underdeveloped or developed countries. In India, Nehruvian socialism had transformed the state into a regulatory body with significant socio-political and financial powers and responsibilities (Das, 2002/2012; Sarker, 2014). However, with the coming of the neoliberal reforms in the late 1980s, which were then institutionalised in 1991, the state began to lose hold of the power that it had once held in the lives of the citizens. Most of the changes that neoliberalism

---

[6]Around 45 percent of Assam's Gross State Domestic Product (GSDP) comes from agriculture (see https://des.assam.gov.in/portlets/state-income). The agrarian economy in the region employs close to 75 percent of the state's working population (see https://asrlms.assam.gov.in/as/node/90627) (accessed: 15.05.2024).

Fig. 8. A Photo From Public School Road During the 2022 Floods. *Photo credit*: Author.

brought forward were in synchronisation with David Harvey's classical yet wide-ranging definition of neoliberalism:

> Neoliberalism is in the first instance a theory of political economic practices that proposes that human well-being can best be advanced by liberating individual entrepreneurial freedoms and skills within an institutional framework characterised by strong private property rights, free markets and free trade. The role of the state is to create and preserve an institutional framework appropriate to such practices ... Furthermore, if markets do not exist (in areas such as land, water, education, health care, social security, or environmental pollution) then they must be created, by state action if necessary. But beyond these tasks the state should not venture. (Harvey, 2005, p. 2)

Neoliberalism in India came about through a replacement of the Nehruvian welfare state that strongly relied on public sector and welfare measures undertaken by the state. The welfare state,[7] despite the shortcomings that it comes to possess, is still an important part of the developing nation because of the role that it comes to play in ensuring social justice, albeit with its own limitations. The most critical role in this regard is played by the public sector, which ensures the fulfilment of the state's duties and responsibilities towards the citizens and also ensures that the benefits and welfare measures enacted by the state reach the citizens in an effective and timely manner. Under a welfare state, the public sector ensures that the management of the state and the society runs in accordance with democratic principles ensuring social justice and equality (Desai, 1984). However, the public sector becomes a major bone of contention between the state and the emerging market, the most crucial and important part of the neoliberal social restructuring.

The market and its associated dynamics often make it impossible for the state to act in the way that it is supposed to act during times of distress, i.e. as a force of welfare provisioning. Neoliberalism, instead, focuses on only the ability of policing that the welfare state possesses out of its wide range of characteristic features and transforms the same into a structure that merely acts at its behest. Under normal circumstances, the need of the state for citizens living under neoliberalism has decreased considerably. However, natural disasters are not normal circumstances. During natural disasters, the people look for solutions that go beyond the narrow consumerism promoted by neoliberal capitalism and that is the point when the relevance of the state – not necessarily as a form of authority but as an elected body of officials – becomes relevant to many people. However, as Harvey (2005) notes, the role of the state under neoliberalism has largely been rendered obsolete, except in certain specific market-friendly situations. Contemporary neoliberal capitalism has reduced the state to satisfy its market-friendly policies. However, it still uses the state to further its influence in the society because of the all-encompassing nature of the state.

The reduction of the role of the welfare state and the increasing domination of globalised production mechanisms has forced both public and private to employ contractual workers to keep up with global capitalist development. Contractualisation has become a necessary part of their strive for profitability because contractual workers help the companies to lower the costs associated with producing and providing a service or a commodity (Deb Roy, 2024b) The inclusion of contractual workers within the various public sector enterprises has also great implications for the services that they are able to provide within disaster-struck societies. During the 2022 floods, the implications of these measures were felt

---

[7]The welfare state can be defined as a state that provides its citizens with certain welfare provisions, which include social security and planning (Desai, 1969). The major objective of any welfare state remains the protection of people from poverty, natural disasters, and the creation of employment opportunities that provide dignified living and fair wages.

30 'Natural' Disasters and Everyday Lives

acutely. The electricity of almost the entire town was shut off during the course of the floods, i.e. from 20 June 2022 until around 30 June and even until 2 or 3 July for some areas.[8] With people making frantic calls (whenever they got some mobile connectivity)[9] and visits to the Assam Power Distribution Corporation Limited (APDCL), the workers of APDCL probably faced the busiest time of their entire lives, as a worker – an electronics engineer by training – who works as a cashier in APDCL says:

> We cannot do anything. There is no fund. Plus, most of the transformer are under water, it is dangerous to go there. We could do with few upgrades which would allow us to touch the transformers and poles, probably then we can provide certain areas with fixed time by boosting them up. But as of now, we cannot do that. Even after the water recedes, it will take a lot of time to get the things in perfect order.

The electricity sector in Assam has been at the centre of major disinvestment plans for many years now.[10] Advocates of disinvestment and privatisation usually argue that privatised services would be cheaper and more competitive than the ones in the public sector, even if they are more unequal by design (McDonald & Ruiters, 2005). Many scholars such as Boardman and Vining (1989) and La Porta and de Silanes (1999) argued that the private sector performs better both economically and socially because it is usually characterised by lesser political manoeuvring, high salary hikes for the employees, and performance-based incentives. However, these claims have found themselves to be debatable within contested terrains because the private firms have exhibited neither the profitability nor the increase in accessibility that had been expected of them (Sathye, 2005; Singh, 2015). During the floods, this failure of the reforms had been quite blatantly exposed with most of the people being left at the mercy of their fates with the state unable to care for them.

In 2022, regions such as Shalchapra, Madhurbond, Aulia Bajar, Koratigram, Kodomtola (as shown in Fig. 9), and Tukurgram – areas outside the mainstream peri-urban region of Silchar and populated by a significant Muslim populace – had to wait for days before the state could intervene effectively in rescuing them from the clutches of the gushing waters. This time as well, the condition was no different, with residents from Tukurgram, Koratigram, and Kodomtola being left

---

[8]Some areas such as Ashram Road and Madhurband and College Road got reconnected after more than five days of the general re-electrification of the town.

[9]It is important to mention that mobile connectivity was disconnected for around 10 days for most of the people of Silchar due to the mobile towers being malfunctional or shut down.

[10]See the news report https://www.pratidintime.com/latest-assam-news-breaking-news-assam/assam-govt-to-privatise-aseb (accessed: 15.05.2024).

Fig. 9. River Water Flowing Into Kodomtola During the 2024 Floods. *Photo credit*: Author.

unattended for days before adequate help reached them. A Muslim resident living in one of the lanes of Kodomtola stated in 2024:

> It has been two days since water is entering the region with high speed, but even now, there is no help from the authorities. There are sick people inside the lane who need help in being rescued out because water might enter their houses anytime if it has not already entered. The government hospital has no nurse on duty, or they are not telling us because we know most of these people are contractual workers who do not want to work during floods. The people are so poor that they cannot afford those fancy private hospitals which are there nearby.

The growth of the private healthcare and contractualisation which became a major cause of concern for these individuals are the results of the neoliberal

32   'Natural' Disasters and Everyday Lives

restructuring of the Indian economy that have achieved a crucial function for neoliberal capitalism, which is a gradual weakening of the public services, especially the service sector. The service sector is an extremely critical aspect of any disaster situation because it is often entrusted with the responsibility of providing essential commodities to the people. The takeover of the service sector by private players represents a failure of the very idea of the welfare state, one which reflects itself quite gravely during a major natural disaster in the form of reduced welfare schemes and provisioning that leaves a significant section of the populace completely unattended during a disaster. The failure of the welfare state structure in India marks a bitter loss for the Nehruvian Socialist era of Indian politics. The dismantling of the welfare state structure in India has meant that the marginalised have been completely rendered powerless in light of the effects that the impending climate change produces for the everyday life of the marginalised populace. The welfare state, in societies such as India, is more critical because the state is the only body in these regions that possesses the ability to provide essential commodities and services to most people because of the region's distance from major administrative centres and the difficult geographical terrain that characterise it.

The state under neoliberalism has to function within a context where its services have often been completely overlooked and deemed unsatisfactory to a section of the people, especially the middle class. Most of the middle class – the class which often oversees the functioning of the society under neoliberalism – has come to harbour an antagonism and hostility towards public services which has made it generate a sense of distance from being engaged in the public service and welfare procedures in place. The middle class have often voluntarily been the vanguards for privately owned services and commodity production, more so during disasters as one of them have stated:

> The government is inefficient. These things should be given to the private companies. Private companies would make their workers work in the floods also. The government cannot do that because it still depends on them for votes. That is why in western countries, you do not see these things happening.

The claim made by the respondent above is a testimony to Mark Fisher's (2009) idea that it is much easier today to imagine an apocalyptic scenario for human survival than to imagine the end of the capitalist organisation of the society. The advocacy of public services as being inefficient is an important aspect of neoliberal capitalism, with advocates of privatisation arguing that under private regulation, all kinds of service delivery would improve qualitatively, with better funds and revenue systems available to them. The opposers of privatisation, on the other hand, bring forward the inherently unequal nature of privatised services which include low wages, unaffordable costs, lay-offs, practising poor health and safety standards, and the like (McDonald & Ruiters, 2005). The perception regarding public services, however, is a class issue that remains contingent upon the ways in which people define accessibility. For some, accessibility is more often than not determined by convenience, while for others, accessibility at a cheap

Disastrous Outcomes of Repressive Monuments   33

price becomes a question of survival. Two individuals – one from the middle class and the other from the working class, respectively – reflected upon their diverse attitudes in these words:

> I have been trying to get an LPG cylinder[11] for days now. It has been three days since the electricity went off. I am trying to get it for so long, but they say that none of their delivery people are available. Without an LPG, how will we cook? It would have been better if we could have a better private service. You of course pay more but then you also get the service more efficiently.

> The food has become scarce. There has been a massive scarcity of food in our home, especially with my husband out of work. There is hardly a few days' food left in the stock at the relief camp, after that I do not know what I would do. I have three children and most of them are waiting for the government to come and rescue us from hunger.

The two statements above made by individuals from very different class positions point towards the class nature of the Indian society and the ways in which class dictates one's positions towards issues and one's modes of articulating the problems that one faces during natural disasters. For the middle class, as the statement above reveals, the government continues to remain merely an entity that is characterised by political corruption. Their hopes rest on the growth of private service provisioning as being more capable during times of disasters. The argument from the middle class is laid on the ground that private services can better manage their workers – because the workers under private regulation are often stripped off their human rights – which makes it better suited to provide the middle classes with the seamless services that are crucial to their class consciousness as the middle class. For instance, the disconnection of water supply was a major issue during the 2024 floods, because the pump that was used to supply water had been damaged, and some people from the water supply department pointed out that they would be unable to get it repaired because of the lack of available funds and workers.[12]

The middle class plays a key role in creating avenues for the deeper penetration of neoliberal governmentality among the poorest of the poor. During natural disasters, the class-allegiances shown by the middle classes become very important because the middle class still controls the various processes through which the state intervenes into the society. For example, many non-governmental organisations (NGOs) – most of them being controlled by middle class individuals or by politically connected people – working at the behest of the state took to distributing

---

[11]LPG refers to liquid petroleum gas cylinder, the most common heating agent used for cooking in urban areas in India.

[12]Statement by a public sector worker, who requested to withhold details.

34   'Natural' Disasters and Everyday Lives

the relief materials, for which they depended on the middle class, both financially and materially. These NGOs during the 2022 floods received a significant amount of donations from the middle classes which contributed greatly to their financial profiles. An NGO volunteer who had worked with an NGO during the floods later recounted:

> I do not know from where they were getting the money, but they had some serious money. Most of them were also promised political gains. There are some good people, who work in them, there is no doubt in that. But most of the people are corrupt and they are in that just for the political gains or the position that they can hold in the community later on.

Most of these activists come from the middle classes, people with some form of social and financial capital. The middle class remains better equipped financially and socially to counter the effects that a flood produces. The middle class, however, is not a politically neutral class and has been a dominant factor in Indian politics and has played a crucial role in the domination that the far-right exhibits in contemporary India (Jaffrelot, 2008). During and after the 2022 floods as well, this characteristic of the middle class was visible to the society. The people from the ruling political formation – the Bhartiya Janta Party (BJP) – became the people who went on to control the entire relief economy during the floods. They were also the people who, in the absence of the municipalities, were given a free reign within certain neighbourhoods to distribute relief materials and attend to local issues as they deem fit. In most cases, the people closest to them – their family, relatives, and acquaintances and then the supporters of the political parties they were affiliated to – were helped in the beginning. This attitude significantly delayed critical support to many vulnerable families who needed it more than others.

The middle classes engage in political processes by focusing on various socio-economic temporalities and cultural loyalties and not purely based on the class that it belongs to (Clark & Lipset, 1991; Goldthorpe, 1995). The implications of these processes have become further grave with the middle-class tilt towards right-wing politics, which is its characteristic feature globally because of the privileges that are often guaranteed to it by the right wing (Burris, 1986). The implications of these processes are evident during a natural disaster and the ways in which the middle class reacts to it which are dominated by a sense of hostility towards those below them in the social hierarchy, a structurally violent attitude towards protecting their own privileges, and a drive for self-preservation. Most of these characteristics make the middle class harbour a dehumanising tendency that is explicit in the way in which they treat the working class, as a domestic worker stated during interviews after the floods:

> I do work at three houses. Two of them did not deduct my wages, but one did. That lady called and said that since I did not work for eleven days, I should not demand money. I told her that I was ready to come as I was staying in a relief camp, but that it was

her problem that her area was under water. She curtly told me to shut up or else I will lose my job. I still work there but the days are gone when I used to think of them as people who would help me in times of need. I see them only as employers. These people got all the relief materials during the floods, that was distributed by the local aides of the political figures, while we in the relief camps struggled for every ounce of rice. Now I see them speaking about the horrors of the floods, and I smile. They do not even know the reality of the floods.

Individuals from marginalised communities do not possess the requisite socio-political capabilities to make their voices heard within a bourgeois democracy, which not only stifles their political assertion but also dehumanises them by marginalising their experience of a particular natural disaster. For them, the public sector and the avenues created by the government and the state become a vehicle through which they assert themselves onto the state and within the society. The middle classes possess the necessary characteristics required to construct the state or the market in its own image and ensure that they know of the issues that the middle classes face (Le Grand & Winter, 1986). For a marginalised individual, the difficulties that one has to endure in making one's issues heard are central to their definition of being a functioning part of a democratic society. In the pre-neoliberal India, the public sector played a crucial role in that regard by providing the citizens with a sense of empowerment because it gave them a sense of collective ownership and control (Ram, 2014). The growth of neoliberalism, however, has weakened the public sector and has also simultaneously weakened the community structures in place that used to serve as the agents of the Public Sector Enterprises (PSEs) and the state during times of disasters, as a retired public sector worker now working as a community activist in Ashram Road[13] states:

Earlier, you could have two three good people in the neighbourhood whom you could tell to help others and believe that they will do their job. Now, that is not going away, but the problem is that most of these people today have become politically motivated and expect that if they do something, they should receive something, either in the form of money or in the form of some privilege.

Contemporary capitalism dictates the entire social reality that an individual lives through, and natural disasters and its associated aftershocks are a part of the same social order. The state under these circumstances is merely a cog within the broad corpus of power relations and exploitative social relations that neoliberal capitalism produces and sustains. The powerlessness of the state has a grave impact on societies such as the Barak Valley because such societies harbour a

---

[13] An area dominated by Dalits in the town, most of them working as rickshaw pullers, pushcart workers, and domestic workers.

36   'Natural' Disasters and Everyday Lives

multiplicity of marginalities – both at the macro level and at the micro level – which often tend to get further exploited within the system created by the market.

Misplaced Trusts and the State

Under neoliberalism, the state has, in most cases, delinked itself from the people and has resorted to acting as a complementary agency to the market. David Harvey (2005) notes that under neoliberalism, the state under most circumstances has become an institution acting at the behest of the market promoting characteristics that do not conform to the classical notions of the state but instead makes it part of a larger framework within which neoliberalism functions. However, during a major disaster, the need of the state becomes extremely important because the state remains the most critical element within the framework through which the marginalised people can exercise a right to the city that they belong to. If the state acts at the behest of the people, it can prove immensely helpful in mitigating the effects that a natural disaster of the scale of the 2022 Barak Valley floods have had, as a social activist associated with an NGO stated:

> The State needed to come in much earlier. They come in when the situation gets out of hand. What is the point in that? They need to come in when everybody talks about the failure of the infrastructure in the region. Why was the [Bethukandi] dam not repaired in advance? Why was it so delayed that it could not even be completed before the second wave of floods came up?

These questions raised by the activist are pertinent ones, especially when, as of 2024 January, 'The dam has still not been repaired, it's just some basic repairs being done there.[14] If it rains for more than three or four days, I guarantee you, we are going to have a flood again, perhaps bigger'. Anxieties raised by these issues have been a cause of major discussions in the region since 2022, as another activist from one of the prominent left-wing political groups in the region notes:

> The Assam government works for the Assamese. They do not care much about us. We are not the Brahmaputra Valley; we are the Bengalis of Barak Valley. Nobody cares about us. Had this been at the Brahmaputra valley, this would have been repaired in a matter of days, and here we are after two complete years, and the place is as it is. I can give you in writing that if it rains like it did in 2022, we will again have massive floods, perhaps bigger.

The statement proved to be true in 2024, merely two years after the 2022 deluge, when again the same Bethukandi dike was the prime reason behind the

---

[14]The dam referred to here is actually a dike, in Bethukandi, the collapse of which was perceived to be the main cause of the greater intensity of the floods.

flooding, and the government again issued a similar statement that the dam had been damaged intentionally by people and requires repair. While it is true that the repairs were done eventually – which many suspect is only temporary because it was done by inserting an iron structure into the sluice gate – but it was only after a significant amount of water had already entered into the city through the cracks. The government had to declare a clampdown around the region as well, something that it had done during the 2022 floods and in 2023 as well during the monsoons.[15]

The apathy shown by the governments towards the region has a disproportionate effect on different sections of the population. While the relatively well-off usually analyse that such lacklustre attitude can be countered by financial prowess, the marginalised have to bear the major effects of the same in the forms of being displaced and left bereft of the basic essentials of life. Within the everyday life, the cities of Silchar, Karimganj, and Hailakandi make it extremely difficult for people coming from the marginalised sections of the populace to make themselves at home within the city. People from lower caste communities, women, and persons with disability face numerous issues – ranging from direct exploitation such as uneven income levels and indirect modes of exploitations such as those emanating from various forms of structural violence. Most of these processes ensure that the marginalised individuals and communities fail to become an effective part of the urban space itself. Citizenship within an urban space, as Merrifield (2014) notes, is always a territorial issue, one which simultaneously includes as well as excludes individuals from being active members based on the class or social groups that they belong to. The propensity of being excluded, however, increases manifold if one comes from marginalised and vulnerable social backgrounds.

The exclusion of marginalised individuals increases manifold in highly underdeveloped areas because these areas remain plagued by extremely low financial resources which are often disproportionately balanced in favour of the elite and the relatively well-off. Barak Valley, because of its underdevelopment, remains dependent on the state government for fulfilling its everyday essential developmental and human welfare-related practices. Without the state acting as a vanguard, it was no surprise that the region immediately succumbed to chaos once the floods had achieved a scale much larger than how it had been presupposed by the local authorities. The Cachar district – the major victim of the floods – received rainfall that far exceeded the records, as has been stated in the Introduction. Despite this, the district and state authorities did not take any preventive measures for the possible floods, knowing full well the dilapidated condition of the dams around the city. The people kept waiting for the state or central governments to intervene, which came in only after the floods had rampaged through more than half of the city. Two residents who live near the dams and had been

---

[15]See https://www.barakbulletin.com/en_US/suspecting-intentional-cutting-of-bethukandi-dyke-cachar-dc-enforces-restrictions-under-crpc-144/ and https://www.barakbulletin.com/en_US/district-administration-clamps-144-around-bethukandi-dyke-no-assembly-of-more-than-04four-persons-should-be-allowed/ (accessed 04.06.2024).

38 'Natural' Disasters and Everyday Lives

displaced during the initial few days of the first wave of the 2022 Silchar floods stated:

> Nobody came to see what the hell was wrong. It was only after the cries of flood were heard that officials came. There was no proactive approach from the government. It was only reacting to what it could see. This approach is the major reason behind the floods. If they had repaired it in the beginning when we had complained,[16] nothing of this sort would have happened. The problem is that we do not have services during normal times. When there was news that floods might break anytime, only then the authorities came into action. If they had taken care of the dams and dykes before the rains, none of this would have happened in the first place.

> I knew the floods would happen. They came to Bethukandi and asked about the dyke[17] there, but nobody had the curiosity or will to check. They came and asked whether water was flowing out, and we said yes. We also said that if water keeps on getting out, somebody will do something because it floods our houses temporarily. And then the big thing, the flood, came. And it was not such that they did not know. They knew that the infrastructure was in shambles.

Most of these factors point towards a widespread existence of underdevelopment in the society. The underdevelopment of Silchar, however, is not a surprise. Being a region that is largely known as being a misfit within the broader northeastern region of India owing to its distinct Bengali identity, the region does not receive the equal amount of funds or political attention. At the same time, the region is also not a highly industrial zone and being situated near to the India–Bangladesh border always carries the risk of being a highly sensitive area. The focus that any particular region received under capitalism is determined by the extent of profits that a particular region can provide. The state as a monument in these regions becomes a ground for struggle. As a state, Assam is one of the least developed regions in the country, and the Barak Valley has often been a region that has not attracted much developmental attention because of the long history of the Assamese–Bengali conflict in the region (Gupta, 2009). In such regions, the dominated populace has to struggle with the dominant ones for their share of the

---

[16]Some of the residents said that they had complained about the bad state of the Bethukandi dike before the floods, but that it had not been acted upon. Non confirmation, however, of this complaint could be found.

[17]A dike normally runs along or parallel to a body of water such as a river or a sea, whereas a dam runs across or through a body of water. A dike has water only on one side, whereas a dam has water on both sides. The main purpose of a dike is protecting the land behind it from flooding, whereas a dams' purpose is to retain the water.

attention that they ought to receive from the state. Consequently, even within the dominated regions, the relatively well-off's struggle with regard to the state is different than the one that the marginalised engage in. The relatively well-off struggle against the state on the terms of the benefits that a privatised service offers, while those that are lower than them in the socio-economic hierarchy struggle against the state on the terms of their basic survival, because the state is necessary for their survival. The struggle between the two can be said to be a struggle over the nature of the state and its meaning for different sections of the populace because the 'Struggles over monuments are ultimately struggles over the world we want, mediated by representational problems pertaining to the aesthetic form and content' (Goonewardena, 2019, p. 190). Groups asserting a strong influence within a particular space, are frequently from dominant caste communities. The dialectical relationship between these communities and others, even if they are locals, contributes to the overall nature of the urban space and the community relationships therein. This is critically evident in the struggles that most Dalit and Muslim individuals had to endure during the floods:

> As Dalits, it is very difficult. We do not get direct entry to some of the camps set up. Some of the people do not want to live with us. So, ultimately you have to stay with your own family or some extended relatives even in the relief camps. People from higher castes sometimes also make a queue in front of the water tap earlier, so that they do not have to get water after us.

> Muslim people have been finding it very difficult. More so, after the flood blame came on Muslims. Now we get rebukes quite frequently saying that we are responsible. Nobody to complain to. It increases because most of the people are also frustrated with the lack of resources and the government is doing nothing. Just keep them happy so they will not bother us.

In both the cases, the individuals blame the state for the mismanagement that had occurred during the floods. The blame on the state also manifests a lack of co-ordination between the central government and the regional governments, which has been noted by both Ge (2019) and Viju (2019) in their accounts of flooding in Southern and Western India. Based on federalism and democracy, the Indian economy had been structured to be a mixed one, and it has

> remained so after liberalisation. For, there occurred little privatisation, while the vast public sector continued to exist. [...] At that level of generality, then, not much appeared to have changed. Continuity with the past is therefore one aspect that is apparent in respect of the impact of liberalisation. (Nayar, 2020, p. 134)

The public sector and its workers, during disasters such as floods, are typical examples of this proposition because the public sector remains the most effective

40 'Natural' Disasters and Everyday Lives

tool at the hands of the state to perform its most critical function during a major disaster, which is that of provisioning of essential services, healthcare, and life-sustaining commodities.[18] The growth of private players within these sectors has caused a relative absence of the state in the lives of the citizens, which are often felt acutely during a disaster because most of these are essential life-sustaining services:

> For many of us, the state does not exist in our everyday lives, but it is only during a major natural disaster that we remember that the Assam state exists. Usually, the State gives us a blind eye to all the problems that we face. The highway still remains half done, so is the alternate route that I have been hearing for such a long time.

> Where is the government? It is supposed to take care of us right? Then where the hell it is. For three days, there is no electricity or water here. There is no initiative. I do not have cash with me, no ATMs nearby, where I should get what I need?

The statement represents the core contradiction that the neoliberal state faces under contemporary capitalism in India. It not only represents a structure of authority but also represents for much of the populace as a structure that comes into force whenever the market fails to live up to the expectations that it itself creates. The state and the market work together in the society to create a de-politicised social consensus even though, as Glaser states:

> [We] are not and cannot be post politics. Politics pervades every aspect of public and private life. We couldn't get beyond it if we tried. Politics is in every choice, every obstacle, every act of pulling rank and every gentle nudge. It's in the courtroom and the canteen, the boardroom and the bedroom. (Glaser, 2018, 'Introduction')

The de-politicised social characteristic that most individuals possess makes them impervious to the political nature of natural disasters, which in turn often makes them analyse disasters as some kind of a divine justice, rather than analysing them as being parts of the overall socio-political dispensation in place. Such explanations get accentuated under neoliberalism because neoliberal reforms make it more difficult for the state to make its presence felt for the highly marginalised and vulnerable. The loss of the importance of the state gets manifested within a general decline in social securities and autonomy of the state, which make it all the more convenient for market-based socio-economic policies to make their mark in the society. One of the major contemporary characteristics of the state is that of being a substitutive one, one which only comes into function as and when the market requires it to function (Berry, 2022). Disaster capitalism becomes a reality

---

[18]These are also services that are often typically more profitable than others and thus are more susceptible to being taken over neoliberal forces (Deb Roy, 2024a).

within and through the neoliberal state because the state remains the most powerful weapon at the hands of capital under neoliberalism to wield in countries like India where geographical and socio-economic development is unevenly distributed. Thus, despite the state being rendered powerless by neoliberal market relations, neoliberalism still continues to depend upon the state for *doing its bidding*.

David Harvey argues that under neoliberalism, the state merely functions as an agent of the market encouraging strong private property rights, harsh legal structures, and free-market-friendly institutions (Harvey, 2005). In the Global North, as Naomi Klein (2007) and Anthony Lowenstein (2015) have exhibited, disaster capitalism came upon through the conditions created by free-market capitalism through an obliteration of the role of the state. In countries such as India, however – which have historically had a very strong statist structure until recent times – disaster capitalism cannot be unleashed without the active consent of the state. However, while full-fledged disaster capitalism is difficult to be unleashed in India, the state acting with the market has been an active force in creating a form of disaster capitalism with its own Indian characteristics using structures such as caste, patriarchy, feudalism, and familial values to further extend its influence within the society. This can be witnessed in situations such as relief distribution, rescue operations, and post-flood rehabilitation programmes which are mostly controlled by the dominant community (through the individuals therein) in place, as was reflected upon by Anjum:[19]

> Most of the relief materials were usurped by those who have political connections with the BJP. We only got leftovers. My family and I stayed in our home only as we could not go to a relief camp. We had a terrace where somehow we survived. It will be wrong to say that they did not visit to give relief materials, but it was mostly handled by political people, like the ones that are connected politically. We are CPM[20] supporters, so why will a BJP person give us first? They naturally gave other people in the beginning. At the end, when everybody had received, then they came to us to give us a packet of biscuit and two litres of drinking water.

The existent social order makes it imperative for neoliberal capitalism to provide a certain section of the populace with avenues to prosper, which plays a pivotal role in the creation of the humane face of capitalist development and neoliberalism in developing societies (Karim, 2021). With large-scale privatisation under neoliberalism, the impact of these transformative processes has increased considerably because the welfarist economic policies of the state have been replaced by the technological rationality of the market which has converted the entire social structure into a giant market where human life has as much value as the amount of profits or surplus labour that it can generate. Floods in the Barak

---

[19]Name changed to protect the interviewee.
[20]The short form of the Communist Party of India (Marxist).

42 'Natural' Disasters and Everyday Lives

Valley have time and again exhibited the necessity that publicly owned services continue to possess in countries such as India, especially within its underdeveloped regions. Natural disasters, becoming a catastrophe in such regions, have the capacity to restructure the nature of the society as one knows it there, as a middle-class private firm salesforce worker informs:

> After the floods, things have not been the same. People are more afraid, even a little bit of rain makes them afraid. Even me, I also think that if there are a few days of incessant rainfall, there would be another flood. So, even with a little bit of rain, now, people begin to stock and hoard things like potato, rice, onions, etc. So, what we have is a mass hoarding almost every time whenever there is a little bit more rainfall either in the valley or in the adjoining hills.[21]

Under such circumstances, the state as an agent of social stability and rationality is of paramount importance, because in India, electoral democracy and its role within the governance of underdeveloped areas still remains an important part of one's social experience as a citizen. The state continues to have great power in regulating such practices, either through enforced legal regulations or through its role in generating a sense of confidence within the populace by focusing on its welfarist nature. The community, in this regard, could play a crucial role because the community remains intrinsically connected to the victims during a major disaster. The community has long been understood and analysed as a force of positive social transgression in human history. It has been positioned as a positive force that can overcome the exploitation of marginalised individuals. The structure of the community is one of the most important parts of the social cohesion that one can find in the society, especially in India (Dube, 1990). A financially liberal political framework dictates that social and ideological processes are designed to favour the relatively well-off and the elites making them equipped to access education and other privileges while constantly impoverishing the others (Losurdo, 2006). In countries such as India, the state also has used community structures in place to bring into effect the various structures of control and authority suitable to its own benefit. The capitalist structure of the state ensures that individuals remain within the structural and ideological contours of expectations which are beneficial to capitalism.

Monumental Amalgamations During Disasters

The state and community both remain as indispensable parts of the social psyche in India, with the latter often being the mode through which the state exerts itself within the society. Communities, within sociological research, have always attracted a wide array of opinions, which have analysed communities to be either merely forms of local expression or means of generating a sense

---

[21]In the Barak Valley, it is widely known that whenever there are rainfalls in the Borail range of hills, there are usually heavy rains in the city and its associated plain areas as well.

of one's identity within a space (Lee & Newby, 1983). In the context of India, the state becomes the major structure of authority, a monument in the words of Henri Lefebvre. Monuments or structures of authority play a great role in mediating how overall urban spaces function, playing a significant role in the direction that urbanisation takes dictating the social and individual lives of the citizens' everyday lives. They interfere with 'place-making practices' dictating the nature of everyday lives, whose effects dictate the production and overall characteristics of the places (Byrne & Goodall, 2013; Cresswell, 2004). The Marxist scholar, Henri Lefebvre, wrote about both the merits and demerits of such 'Monuments':

> *Against the monument.* The monument is essentially repressive. It is the seat of an institution (the church, the state, the university). Any space that is organised around the monument is colonised and oppressed ...

> *For the monument.* It is the only conceivable or imaginary site of collective (social) life. It controls people, yes, but does so to bring them together. (Lefebvre, 2003, p. 21, emphasis original)

In his dialectical fashion, Lefebvre lays down before the readers the 'pros and cons' of the existence of monuments. In the words of Lefebvre, the arguments against 'monuments' are appropriate for institutions constructed as symbols of victories by authoritarian individuals, states, or capitalism as an epitome symbolising its oppressive regimes. For any kind of social transformation – that desires to go beyond the logic of capital – the latter kind of monuments become the major obstacles because of the kind of ideological effects that they have on the society in the form of restricting the emancipatory vision of the marginalised to different forms of statist formulations. Both the state and the community in the context of India often act as repressive institutions mostly based on the pre-capitalist institutions of caste, race, gender, and ethnicity. In the Global North, most communities are often merely contrived ones.[22] This is, however, not the case for the SGS. For people within the SGS, the community is something that they become a part of involuntarily, often by birth depending upon one's caste or creed or ethnicity and often have no control over. The community then dictates the ways in which they live their everyday lives. Few of the black marketeers – who sold rice and lentils at higher prices – also recounted how it was their community leaders themselves who encouraged them to do those activities:

> I did not want to do that. But then, X[23] came and told that if I sold one bottle of water I get 10 rupees. It was a very difficult time, so I also complied. The price was set that for 500 ML bottle, the selling

---

[22]Communities formed on the pretext of fulfilling some purpose, like a student city or a working-class industrial neighbourhood.
[23]Name withheld on request.

44 'Natural' Disasters and Everyday Lives

price would be 50 INR. The leader had some sort of arrangement with a stockiest who wanted to get rid of his stock of bottles or something, I am not aware of it.

The state for most of them is an entity that is an abstract one, detached from their everyday lives. However, that being said, the state continues to play an important role because the state reserves for itself – even under neoliberalism – all the necessary powers during times of crisis, be it socio-economic or political. In India, this occurs through the Disaster Management Act which came into force in 2005 after the 2004 Tsunami, that has often been criticised for its lack of clarity with regard to its own internal mechanisms and lacklustre analysis of disasters often resulting in delayed responses and bureaucratic mismanagement (Chaturvedi, 2020; Pande, 2023). State responses to disasters usually are aimed at managing them rather than preventing them. The National Disaster Management Authority (NDMA) directs all the concerned central and state bodies to be prepared for a disaster so as to avoid any procedural delays (Ge, 2019). However, as Ge (2019) argues, the NDMA's directives rarely get the attention they deserve. The state's involvement during disasters often becomes mandatory, especially within the domain of provisioning of essential services and commodities. For example, during the Covid-19 pandemic, the Supreme Court had asked the government to consider fixing a ceiling price for essential Covid-19-related drugs under the Drugs (Price Control) Order, 2013, which allows the government to intervene in the retail pricing of certain drugs during situations of distress. Such steps taken during times of distress, despite being praiseworthy, fail to be permanent fixtures in the way in which the country's political economy is run at the grassroots. These kinds of formal measures rarely get implemented fully at the local level, and are mostly run according to the local norms in place.

The marginalised people of Silchar continue to get impacted by floods because the formal measures enacted to safeguard them do not make an impact at the grassroots. Systems such as the weather and meteorological department constantly sending out messages stating that there would be heavy rainfall and warning people that they should stay away from the riverbanks when the water of the Barak rose only cater to those people who can afford to stay away from them. These processes only help those who are within the formal processes and make no dents within the already marginalised lives of the marginalised populace. This becomes evident in the relief camps where despite being categorised as refugees, people without a national identification card – known as the Aadhar card – do not receive timely relief materials or other services as many relief camp dwellers – especially the women among them – stated both during the 2022 and 2024 floods. During the 2024 floods, one such instance was pointed out by the mother of a two-year-old child – shown in Fig. 10 – who injured himself while evacuating his house located near the river. The mother failed to secure government-sponsored medicines for two days because she did not have the adequate documents. She also did not have money to buy medicines from retail outlets because all her money had been washed away. This points towards the distinction that exists between

Fig. 10. A Child Whose Hand Had Been Fractured While Being Evacuated By His Mother During the 2024 Floods. The Mother Had to Wait Two Days for Being Able to Obtain Adequate Medicines Due to the Floods. *Photo credit*: Author.

formal and material equality in the society, which had been one of the basic ideas proposed by Marx who had suggested that:

> [The] rights of human beings are in reality the rights of bourgeois men in civil society. They protect individual capitalists in their exploitative practices, and they protect the unequal economic results of such practices. Rights are associated with individuals who 'own' them in order to protect private interests. Rights thus shield the basic inequalities and exploitative practices of bourgeois culture. Bourgeois culture ignores material inequalities and slavishly adheres to formal legal, moral or political equality of rights. (Vincent, 1993, p. 385)

'Natural' disasters expose the most inhuman aspects of humanity for the world to see and experience. The ignorance of material inequalities and its subsumption

46 'Natural' Disasters and Everyday Lives

within a more formal structure enables capitalism to function as a social structure that produces a kind of alienated empowerment for a certain section of the populace. The disaster management plan brought forward by the Cachar district authority also became embodied by such issues. The plan's focus remains on the steps that the government can take at the bureaucratic level and does not focus enough on the community except the engagement of community volunteers, NGO activation and *Project Suraksha*,[24] which remains more connected to the issues that flood victims face than the bureaucracy. Like the Chennai floods, as Ge (2019) narrates, in Silchar as well, the consensus among the community leaders and local politicians after the floods was that they had been unable to cater to the needs of the marginalised and the victims because they were waiting for orders from higher authorities. However, the broader community that people become a part of often becomes the basis of their exploitation in India because the community is also formed by the established power dynamics in any society that favours the dominant community in place (Viju, 2019). Ge (2019) in her book on Chennai floods has noted the importance of the National Disaster Response Force (NDRF) during the floods. During the Silchar floods of 2022, the NDRF played a critical role as well. The contributions that the 28 NDRF, along with the SDRF and other armed and paramilitary forces (District Disaster Management Authority Cachar (DDMAC), 2022a), had made have also been acknowledged by many of the survivors and victims of the floods. One such individual stated:

> The NDRF people came on time, and we respect that, but the problem is not that they could not come on time, but that the way in which they were directed to do their duties. They came alone, and were unaware of the geography or the locations that people were stuck in. Most of them came to the ward commissioners, and they promptly directed them to people who they knew personally, then they sent them to people who were from their own caste and then we came.

Despite acknowledging the contributions that these forces had made; the survivor also recounts the highly discriminatory nature of the work that the NDRF and others had done. This again brings one to consider the fact that the community in India is a crucial part of the Indian social structure (Dube, 1990), one which determines the ways in which a particular area functions or the ways in which an area is being served by the state. However, as neoliberalism progressed, the community has slowly been robbed off all the welfarist aspects that stemmed from its once autonomous nature such as relief provisioning and welfarist decision-making capabilities, making the people completely dependent on the state during times of natural disasters. As a community leader stated:

---

[24]'Project Suraksha' was 'initiated by District Administration in coordination with Impact Weaver to help out the distressed and connect with the people in need' (DDMAC, 2022a, p. 14).

What can I do? There are issues. Today, with the present government, I do not have a choice. It is either their way or the highway. If I do something different, then I am done. I will have to protect myself right? People do not understand that. They come and say we do not have relief materials and we want things to survive, am I running a bank? Where will I get so much money or things?

It is worthwhile to mention that similar issues continued to function in 2024 as well, with many local activists and community leaders expressing helplessness in addressing the concerns of the people. The dependency that the people have on the state for survival in underdeveloped regions constitutes the basic nature of the society in the SGS. Structures of domination and authority that come into being in the SGS during natural disasters are characterised by a combination of the state and the community. The kind of devastations that floods cause, as Deshpande (2018) has noted, often requires interventions that go beyond the regional government or the urban local bodies (ULBs). Both these kinds of bodies fail to become effective agents of social regulation during a major natural disaster because both of them operate on behalf of the ruling disposition in the society.

Along with the NDRF, the other example that can be cited in this regard is that of the distribution of relief materials – usually comprising some basic food items, over-the-counter medicines, and drinking water – which were also unevenly distributed among different communities depending upon the social composition of a particular neighbourhood. For example, in the upscale neighbourhoods of Ambicapatty or Central Road, the 500-mL water bottles were sold for around 50–60 rupees depending upon the situation – whether the seller had to deliver the bottle or whether the buyer came to buy it along with the age or gender of the buyer – one such post from social media is shown in Fig. 11 – and the scarcity faced by the seller. The state and government remained unable to cater to those in need which made them look for such opportunities.

In India, the state as an overarching structure has played a dominant role, right from the ancient monarchies to the contemporary times of neoliberal capitalism. It has not only affected the society economically but also at a fundamental level psychologically affecting the ways in which people think and react. For example, the people of the Barak Valley, especially from the hinterlands, have always

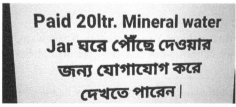

Fig. 11. A Poster Advertising Paid Mineral Water Bottles for Being Sold at Heightened Prices on Social Media During the Initial Days of Flooding. *Photo credit*: Author.

48 *'Natural' Disasters and Everyday Lives*

found the state at a distance from them. For most of the marginalised populace – the Dalits, the Muslims, and women among others – the state becomes visible only during elections and is completely absent during times when they need it the most. As one Dalit flood victim in a relief-camp stated:

> They will only come to us expressing their sadness and emotions during elections. After elections these people go just absent. They did not even visit us during the flood. They came and put up a poster saying that the member of Legislative Assembly (MLA) is here with us[25] and sympathise with us, but where is the MLA?

One can notice a striking similarity of the statement made by the above individual in 2022 with the statement reproduced below by a Dalit woman in a relief camp during the 2024 floods:

> It has been three days now that we are here. Only the school[26] authorities have come but the local ward commissioner and the MLA or Member of Parliament (MP) have not yet visited us. But some of their party-people have come and put up a few posters that they are with us. Maybe, they are with us, who knows?

A key reason for the absence of the state from the everyday lives of the people is the erosion of the public welfare services. After neoliberalism, the character of the state underwent a significant change. Instead of analysing individuals as citizens towards whom the state had an intrinsic responsibility, it began to classify them as being consumers. This transformation of the state also necessitated that the state adopts a highly exploitative attitude towards the marginalised populace. Most of these functions are accomplished through constant privatisation and rendering powerless the erstwhile state monopolies, the effects of which are felt especially in sectors which are deemed to be essential for life-sustaining activities in contemporary societies such as water supply, electricity distribution, and telecommunications. Under a political structure which aids in the construction of a competitive market, state monopolies inadvertently pave the path for the coming of full-fledged neoliberal capitalism leaving the public sector with neither the financial nor the political autonomy required to act in crisis situations. This is something that many of the essential public sector workers had recounted once the floods had been over:

> We cannot do anything. You can beat us if you want, but the reality is that we cannot work like this. We do not have that much

---

[25]Most relief camps and major junctions in the city had large hoardings of the local politicians from the ruling political formations exhibiting their 'solidarity and good wishes' for the people.
[26]The relief camp was set up at a school.

money to do all of these at such a short period. It will take a long time to get things back to how they were. We do not have enough equipment, and neither do we have sufficient finances to do that. Moreover, most of the workers who will actually work to repair these things are all contractual and they said they will not come during the floods as their houses are inundated.

This was a statement given by an electricity worker upon being confronted by a group of 3–5 young people who went on to threaten the worker regarding the absence of electricity. A member of the group later recounted:

I would not have said it on any other day, but that particular day something snapped. It was hot and humid and water all around for the good part of the last ten to twelve days. Then we see this person coming and just checking and go, that is when we got angered. It was so frustrating; these people eat out our tax money and then do nothing.

In India, the state has played a dominant role, right from the ancient times to the contemporary times. After neoliberalism, however, the character of the state underwent a significant change. Instead of analysing working individuals as citizens towards whom the state had an intrinsic responsibility, it began to classify them as being consumers. This transformation of the state also necessitated that the state adopts a greater authoritarian role engaging more intimately with the market processes. In the case of India, it directly becomes a part of the free-market propaganda as the Indian state has been an important part of the developmental trajectory in all of South Asia. In countries such as Sri Lanka and Bangladesh, the Indian state has provided great support to the economic and human developmental plans and programmes (Bhasin, 2008). The hyper-vanguardist model of the Indian state in these regions, however, has not been reproduced in its own hinterlands. India's hinterlands have been left abandoned by the state to the markets for them to make profits from, either through large-scale extraction of minerals – such as Odisha – or by using these spaces as sites from which cheap labour power can be extracted, such as the districts of the Barak Valley from which scores of migrant workers flock to cities like Bengaluru and Gurugram to work as precariat workers. The lack of focus on these spaces have resulted in many of these spaces suffering from developmental lacunae which have made them ideal spaces for global capitalism to penetrate into.

At the local level, the state and the community remain indistinguishable in much of the Indian society with both of them acting at the behest of one another in general. This is most explicitly visible in the case of the informal workers, who work on a daily basis in small towns. There are innumerable reports about the state of informal workers in the country. The Arjun Sengupta Committee Report of 2007 had stated that more than 90% of India's workforce is engaged

50 'Natural' Disasters and Everyday Lives

in informal work.[27] In Silchar, most of these workers remain restricted to a few of the neighbourhoods, primarily within neighbourhoods such as Ashram Road, Harijan Colony, Tarapur, and Kalibari Chor.[28] The informal workers remain bound by local laws, which are often different from the regular legal ones. Unlike the formal workers, these workers did not get any paid leave, for them, the floods, as one of them put it, 'meant that we lost twenty days of work, some 8,000 to 12,000 INR rupees, and along with that destruction of our lives'.

The informal workers also form one of the most important parts of the ecological governance in the Barak Valley because it is – at the end of the day – the informal workers who interact directly with the environment on an everyday basis. They form the core part of the social fabric in areas which get affected and affect the flood dynamics the most in the region. But these are the very communities which have a much more structural and intimate relationship with the environment because of their social and intimate connection with the land that they inhabit (Sharma, 2017). Dalits, experts in traversing the Barak, become the saviours of the city. Unlike other disasters, where their services are *mostly* paid for meagrely and are received without little acknowledgement of the structural violence meted out to them, the importance of their labour increases manifold during the floods in the area. During the Silchar floods of 2022 as well, the Dalits of the city – the Kaibartas – took it upon themselves to provide the city dwellers with a means of transportation. Most of the once busy roads – the Central Road, Hospital Road, College Road, Club Road, and Park Road[29] – were filled with boats being manned by them, with them charging anywhere from INR 50 to INR 200 for a ride that ranged from one to three kilometres. These boats – one of which is shown in Fig. 12[30] – substituted the autorickshaws in the town and took it upon themselves to make the inhabitants of the town able to transport themselves, their families, and belongingness. However, these activities were not taken well by some of the inhabitants. A middle-class individual stated:

> These people are looting us. They are all Kaibartas, you cannot talk with them properly, they will curse you with the worst words that they know. Simple solution is just give them whatever they ask, fifty rupees, hundred rupees. You cannot bargain with them

---

[27]The full report can be accessed through the official website here: https://dcmsme.gov.in/condition_of_workers_sep_2007.pdf (accessed: 15.05.2024).

[28]All these areas are located within a walking distance from important socio-historical, economic, and political sites in the city and remain highly congested sites with a high population density.

[29]There are major roads in the city. They do not have much housing capacities but rather serve as major business streets in the city. Central Road serves as the major road running through the centre of the town, while Hospital Road has the Civil Hospital of the town. Park Road and Club Roads is one of the major centres that house most of the bureaucratic and governmental offices.

[30]The faces have been blurred in the image due to privacy concerns.

Fig. 12. A Boat in Operation During the 2022 Floods. *Photo credit*: Author.

because if you say anything, they will quarrel with you. I paid INR 2000 for a 3 km ride in the water to see my under construction house. They are just thieves.

For the marginalised, such experiences are nothing new. For such individuals, the community during natural disasters such as floods becomes the basis of the ways in which individuals cope with the changing contours of their everyday lives. A community helps most individuals in coping with the changes that such disasters and their associated processes bring in within the society. However, that is not the case for marginalised individuals. For most marginalised individuals, the community itself becomes a part and parcel of the exploitation that is meted out to them on an everyday basis. Acting as a repressive monument on behalf of the state, it negotiates on its behalf, with the various forms of modernity as means of navigating through their everyday lives. Giddens (1991, p. 4) refers to modernity as an entity or a historical form, which reduces the risks in certain aspects of life while simultaneously increasing risks in some other aspects of life. For a middle-class individual, the community becomes a space where one can seek help, while for the marginalised, it is the very institution that constitutes a hegemonical control over them.

The generation of hegemonical control and domination is not a spontaneous process because 'hegemony [on the other hand] is not simply deceit [but is] built around channeling genuine social contradictions in a manner that supports the continued dominance of the ruling bloc' (Gopalakrishnan, 2009, p. 13). This is most aptly visible in the context of the informal workers, for whom the social contract in place works differently from the one that exists for the workers in relatively organised and formal industries. The informal workers suffer from the distinctions which exist between the ideal and real conditions of their employment

52 'Natural' Disasters and Everyday Lives

terms (Plagerson et al., 2022). The implications of such differences are visible across the social spectrum, most notably among Dalits, Muslims, and Women in India.[31] The marginalised suffer from a lack of access to material resources and as such also suffer from a lack of the necessary structural resources that they need in order to become capable of making themselves access those resources.

Amitav Ghosh (2021), the acclaimed novelist, argues that while it is possible for development to occur along the lines of sustainability, such politics of development is rarely found within contemporary political structures and societies. Contemporary capitalism, even within the Global South, has been a key force in the ushering in of ecological destruction of highly sensitive ecological zones, which have increased the propensity of natural disasters occurring across the globe. It is the grandeur associated with natural disasters that often makes them transgress the boundaries of class, caste, gender, and race in societies such as India. They evoke a sense of oneness in the society, as Ramesh (2019) argues. But, at the same time, it is essential to realise that individuals from marginalised populace have always been at the risk of being affected disproportionately by disasters. The ways in which the vulnerable and the marginalised are treated within a disaster speaks of the general conditions of marginality that they face in the society. Floods cause a plethora of issues pertaining to heightened marginality to emerge in the society, which are not only restricted to loss of livelihoods, productive capacities, and lives but go deeper than that creating psychosocial and political effects on the people.

---

[31]For example, for the women, the society conceptualises women's household labour as 'work' which often does not encompass the full spectrum of the efforts that one has to undertake in performing those tasks (Kalpagam, 1994). The definition of 'labour', however, often entails within itself the idea of effort that is not evoked while defining and describing the reproductive labour that women do within the households (Kalpagam, 1994). Another example can be cited regarding the Muslims and Dalits in the Indian society, who have been suffering from disproportionate representations even within sectors which promise to be based on meritocratic principles (Deb Roy, 2024b).

Chapter 2

The Ravages of Relief Activities

When human beings face a crisis, the first reaction that they exhibit is that of fear, a fear of the unknown. Capitalism uses this fear to push forward its agenda among the people in such a way that it generates a state of wilful consensus in favour of capitalism among them. The wilful consensus surrounding the existence of hierarchies within the society presents to capitalism – even in the hinterlands – a scope to bring in further exploitation and hegemonical domination suiting its goals of profit accumulation. Under neoliberal capitalism, the delicate balance that existed between human life and environmental sustenance has been manipulated. It has not only produced a hyper-promethean vision as a normalised imagery but has often made it mandatory for people to engage in the realisation of this vision of modern capitalism. This has been particularly devastating for areas which are adjacent to rivers. During the floods, the people closest to the riverbanks suffer the most, with most of them having to shift overnight to higher grounds or to relief camps – one of which is shown in Fig. 13 – to save themselves from the water.

Bhattacharya's (2022) work on Bengal shows the different aspects of the social realities that people engage with when they live in a low developmental region in India. He shows how people's lives in areas dominated by a riverine system depend upon the ebbs and flows of the river(s). Taking the example of the Sundarbans,[1] Bhattacharya shows that in such areas, rivers determine the nature of the lifeworld that people constitute around themselves. It determines not only the quality of life but also the ways in which people engage with their various economic and social lives. In other words, in an area that is dominated by a riverine system, the river often comes to dictate the everyday lives of the people. In Silchar, the Barak river proves to not only be an ecological and natural entity but also a structure that determines the identity and everyday lives of the people in the region.

For most of the people from the Barak Valley, the river Barak has become an identity for them because of the ways in which the state of Assam has been

---

[1]Sundarbans are a mangrove delta formed by the confluence of the Ganga, the Brahmaputra, and Meghna rivers in the Bay of Bengal. It is easily accessible through both West Bengal and Bangladesh.

'Natural' Disasters and Everyday Lives:
Floods, Climate Justice and Marginalisation in India, 53–78
Copyright © 2024 by Suddhabrata Deb Roy
Published under exclusive licence by Emerald Publishing Limited
doi:10.1108/978-1-83797-853-320241003

Fig. 13. A Relief Camp in Silchar During the 2022 Floods. *Photo credit*: Author.

divided into the Brahmaputra and Barak Valleys, which has had wide-ranging effects because of the struggles focused on culture that has plagued the relationship between both these valleys and has been one of the most important aspects of Assam's political trajectory (Barman, 2023; Deb, 2019). Such struggles and the intimate association with the Barak river provide the key terms for constituting the lifeworld of the people of Silchar. The effects of everyday life characterised by activities such as walking, jogging, eating out, and having a chat at the nearest teashop provide pathways towards integration with the urban space are related to one's 'lifeworld' (Seamon, 1979, 2000). Lifeworld, according to Seamon (2000, pp. 6–7) refers to the 'tacit context, tenor and pace of daily life to which normally people give no reflexive attention … (which includes) both the routine and the unusual, the mundane and the surprising'. Within this framework, habits can be characterised as recurring behavioural tracts often done involuntarily (Seamon, 1979, p. 38).

The ways in which any individual forms 'habits' and 'behaviour' are intimately related to the environmental conditions that one finds oneself in within one's everyday lives. As authors such as Choudhury (2021), Bhattacharya (2022), and Barman (2023) have argued, the everyday life of people who live near rivers exhibit a certain dependency on the river. The people of Silchar, for example, depend upon the Barak river for water, winds, and, most importantly, an identity of their own. People living near the banks of the river depend on it for survival, while those who live at a distance from the river depend on that for fresh air and winds. The contribution that a river makes to a populace settled by it cannot be analysed merely through material terms but also needs to be seen in the context of the effect that it has on the identity of the people living beside it. With the growth of population, there occurs an increase of human activity in such regions – disrupting the natural order, excessive reliance on dams, and increasing developmental waste – with a view to tame the river that directly affects the riverine system creating an immense amount of pressure on it (Cook, 2019). These activities increase the pressure on the river and make them prone to floods, more so in the contemporary times, because of the incessant technological and industrial development that contemporary capitalism renders mainstream. This has produced a grave increase within the kind of risks that characterise the lives of those living beside them affecting human lives economically, socially, culturally, and politically. Steve Matthewman argues:

> Disasters are on the rise, increasing in magnitude, frequency and cost; We are also seeing new forms of disaster emerging in which 'the impossible' happens. These arise from growing interconnectivity and complexity in our world, which link back to questions of political economy and globalisation; Definitions of disasters have remained static in these most rapidly challenging times of all; disaster studies is threatened with intellectual marginality at the very moment of its greatest need; Even in wealthy countries we do not really know the true casualty figures from disasters, which has obvious policy implications, The spectacular events that disaster researchers focus on may not be the ones which take the greatest toll, A sociology of disasters which omits the most devastating types of damage is barely worthy of the name. (Matthewman, 2015, pp. 8–9)

People who live beside rivers get affected by the changes within the riverine structure directly, especially with the coming of floods and other such natural disasters. They remain perennially under the risk of being displaced. During a flood, they are often the first people who end up in the relief camps set up by the government. The current chapter focuses on these relief camps. This chapter takes the readers to the relief camps which had been set up to help those affected by the floods in Silchar. These relief camps had been set up in various schools, colleges, and other educational institutions. Drawing from interviews conducted with individuals residing in the relief camps and the residents in spaces around the relief camps, this chapter brings out the ways in which an alternate economy is often set up in these camps and the processes in which living temporarily under such an economy shapes the subjective selves of the concerned individuals enabling them to engage in various forms of disaster capitalism.

Struggles Within Relief Camps

The river, Barak, flows by the city of Silchar providing the town with the necessary conditions for constructing a civilisation of its own. However, with the growing pressures of urbanisation, the river is facing a crisis. The high sedimentation in the river has made it vulnerable to more frequent floods, changing the nature of the river during the monsoons. While a decade ago, floods used to happen after

56 'Natural' Disasters and Everyday Lives

around 10–12 days of incessant rainfall, the contemporary society faces such risks more regularly. An experienced resident of Annapurna Ghat,[2] Rakesh,[3] who has been a boatman for more than a decade in the area states:

> Earlier, we used to like the rain. One reason was that rainfall made the river so beautiful to look at. There were also some additional factors involved like agriculture and increased ferry prices.[4] But today, most of us hate it. It is like with a little bit more rain, the probability of a flood has gone up tremendously. Today you do not need a month of rainfall, a week will do the job.

For people like Rakesh, a trip to a relief camp has become an annual event because the areas where he lives – around 100 metres from the riverbanks – gets flooded with even a little rainfall. The growth of the probability of floods and its associated displacement makes it important to analyse the internal dynamics of relief camps, which become the primary shelters for most of the people affected by them. In relief camps, the people engaged in different kinds of activities, most of which would not have been looked at kindly upon during the normal times. However, during natural disasters, the people's drive for survival often engulfs their sense of morality because morality is also an aspect of life that is connected to and is contingent upon the urban space within which it is located. Most scholars – coming from the corporate-based technological fields such as Deshpande (2018) – focus on urbanisation, often without much emphasis on the contradictions that urbanisation produces. In doing so, their focus often remains mostly focused on analysing the effects of the problem rather than on the core roots of the problem, which is market-driven urbanisation policies.

The growth of urbanisation in India has caused a massive transformation of the society. It is well-known that urban spaces in the third world are characterised by growing inequality and structural violence (Kassarda & Parnell, 1993). The effects of such violent activities are most acutely visible within the problems that people face with regard to housing. Housing constitutes one of the major problems within disaster-affected areas, especially in Assam, with more than 100,000 people displaced every year due to the floods. Kalita reports in the *Times of India*:

> For several lakh families in Assam, abandoning their flooded homes to stay in schools, namghars (Vaishnavite places of worship), under tarpaulins on highways or close to railway tracks

---

[2]It is one of the ghats located by the Barak. The ghat also serves as an important ferrying point in the region and serves around 300–400 people on an everyday basis, enabling them to transport themselves and their materials or possessions for work or other purposes.

[3]Name changed on request of the interviewee.

[4]Ferry prices generally increase during the monsoon season because of the rains.

year after year has become a way of life. The more unfortunate ones among them get a new identity – landless – because riverbank erosion snatches away what the rain doesn't. In other flooded areas where there is little to no erosion, the land is saved but the homes are gone – washed away or damaged by flood waters. (Kalita, 2022, para 1)

With the absence of any social justice-oriented housing policy globally, the poor and the marginalised remain bereft of any proper governmental housing plans, regulations, and policies. During any major natural disaster, housing constitutes itself to be one of the major concerns. During the 2022 Barak Valley floods, more than 2,000 houses were destroyed (Krishnamurthy, 2022d). The lack of proper housing along with the loss of income-generating avenues makes it necessary for many to depend on the state for survival. The intervention of the state takes place in many ways within disaster-affected areas as the book has and will continue to reveal. But the most critical function of the state takes place in the form of its establishment of relief camps. During any natural disaster which displaces people, relief camps turn out to be the most immediate and necessary part of the response that societies exhibit. This is because they address the most critical parts of the corpus of problems associated with a flood, housing, medical care, and provisioning of essential services during disasters.

Relief camps in flood-affected areas become akin to Noah's ark rescuing people and providing them with a safe space during the floods (Lal, 2019). Relief camps, as this chapter will show, also become one of the most important sites within which the entire disaster economy shapes up in the Barak Valley. This contribution that a relief camp makes is very different from the basic purpose of a relief camp, as is obvious, which is to provide the flood-affected people with a form of temporary shelter and access to essential life-sustaining services and commodities. The most critical of all the functions, however, is that of housing. As Krupa Ge (2019) has narrated, the most difficult choice that individuals often need to make during the floods is related to relocation, because the idea of 'home' does not only signify material objects but also the most critical aspects of one's self-subjectivity and self-identification within a chaotic society. As a flood-displaced woman in her mid-30s narrated after the 2022 floods:

> After the floods, the homes do not remain the way that they used to at one point. They are just different. It is probably because with floods, many things get washed away, and they are not all material things, our memories get washed away. The loss of material things is only the tip of the iceberg. We lose so many memories, so many things that probably hold no material value but immense emotional ones. During the floods, it washed away an old saree of my mother, which I had held dear for so many years. It had no material value, but it cannot be replaced.

58 'Natural' Disasters and Everyday Lives

These are the highly complicated terrains of loss that floods bring with themselves, which are difficult to be expressed in mere quantitative terms. The reduction of human lives to a numeric value is the basic characteristic of neoliberal disaster capitalism (Lowenstein, 2015). Such a quantitative focus cannot grasp the immense emotional and psychological violence that is inflicted upon people during the entire process of rapid relocation that is characterised by a loss of one's emotional self and subjectivity (Thurnheer, 2014). Relief camps serve the medical, sanitary, and housing needs of a flood-affected community (Singh et al., 2016). They can be simply defined as spaces designed to house people temporarily during disasters. According to the National Disaster Management Authority (NDMA):

> Relief shelters and rehabilitation camps shall be set up in order to accommodate people affected by a disaster. The camp shall be temporary in nature, with basic necessities. People shall be encouraged to return to their respective accommodation once the normalcy has been returned. (NDMA, 2019, p. 2)

Relief camps are supposed to provide the victims of a disaster with a sense of security, and that remains the first responsibility of any government during a natural disaster like a flood that displaces a large number of people. In most cases, relief camps get inhabited by people who come from marginalised social backgrounds. The basic structure of a relief camp remains simple, as a relief-camp dweller himself stated during the floods:

> It is quite simple. You get a large space, put some 100-300 people together with an officer or supervisor in there, and you have a relief camp. Most of the governments stop at that, they do not pay much attention to what goes on there once it has been set up. There are widescale issues of discrimination within the camps. People from lower castes and poor people face immense discrimination because you throw all of them to stay together, under a supervisor or a camp officer[5] who is mostly an upper caste and relatively well-to-do person, who hates us.

Relief camps get occupied mostly by people from lower socio-economic backgrounds, because it is them who occupy lands closest to the Barak in the region and as such get their homes inundated by the floods. During the 2022 floods, there were a total of 292 relief camps and 129 relief centres able to accommodate a total of 5,443 people set up across the district (DDMAC, 2022a, 2022b). Most of these relief camps housed people from poor socio-economic backgrounds, mostly

---

[5]The appointment of a supervisor is a mandatory requirement of any relief camp (DDMAC, 2022b).

comprising daily-wage workers, domestic workers, toto or tuk-tuk drivers,[6] sanitation workers, and people from marginalised castes. The conditions that people face in relief camps – the ghettoised conditions of which are portrayed in Fig. 14 – despite the official methods of supervision reflect the crisis that most of the marginalised communities face. The kind of marginality that is evident in the relief camps reflects the society within which the camp is located, more so in a highly unequal society such as India. One such relief-camp dweller, Revati,[7] stated:

> We stayed just by the river; the river was right next to us. My family has been there for decades now, but never in our lives have we seen the Barak waving at us with such ferocity. The river entered our house and completely destroyed everything. My husband's cart is destroyed, whatever we could gather, we brought it here, but after bringing it here as well, there is no respite. You have people whom we need to pay here to ensure that we have access to safe drinking water. There is no maintenance of hygiene.

Her husband, Alok,[8] used to be a teaseller in one of the business areas of the town, earning around INR 200–INR 300 (2 to 3 GBP app.) a day, which was just sufficient enough to sustain their family of four – Revati, her husband, and two

Fig. 14. Inside View of a Relief Camp During the 2024 Floods.
Photo credit: Author.

---

[6]'Toto' and 'Tuk-tuk' are forms of an autorickshaw usually found in South Asia. In Silchar, 'Tuk-tuks' are one of the most widely used means of general transportation mostly operated by the Kaibartas and Muslims.
[7]Name changed on request of the interviewee.
[8]Name changed on request of the interviewee.

60 *'Natural' Disasters and Everyday Lives*

children – but with the floods, the shop was washed away, leaving them with no income-generating avenues. Their relocation to a relief camp at College Road, allowed the family to search for an alternate mode of income during the floods that allowed them to survive. The couple resorted to selling water bottles, a prime necessity for those who continued to stay in their houses. She further says:

> We could not have done anything. We went to a shop to get food, some rice and dal[9] but the prices were sky high. There were saying that we need to pay some fifty rupees for half a kilo, and we have a family of four. How will we do that? We do not have any income also at the moment. No doubt, some of us went onto work as businesspeople. It serves these people right, who see us less than human beings on general days. During floods, they came to depend on us, it was fun actually.

Relief camps are also a part of the society and are formed through the existent governmental practices in place. Within the relief camps as well, caste, religious, and racial segregations continue to be operational. However, the fundamental character of a capitalist society goes beyond such issues. Issues such as caste, race, and religion are important, but as Kevin Anderson (2020) notes, they operate within an overarching capitalist social reality and thus cannot be considered to be completely autonomous from the broader reality that capitalism creates and sustains. The overarching presence of capitalism is felt across the economic, social, and political situations that one finds during disastrous times, most explicitly within the relief camps that come to be characterised by extreme poverty, extremely low sanitation, and little relief materials.

People in the relief camps, mostly devoid of any income during the floods, often have no other option but to get engaged in activities that are looked down upon by the mainstream society. Examples of such activities might be selling certain commodities such as mosquito coils, water bottles, and dry imperishable food items such as biscuits and cakes. During the floods, the relief camps set up by the government in schools, colleges, and other such open spaces function through the same contradictions that one finds in the society, as a relief-camp official and community leader explains:

> It used to be simple a few years back, like people were simpler. We did not have to do much of the arrangement, But today, we have to take stock of what communities they are from, whether they are comfortable with each other, whether they will fight and all that. There is a huge amount of corruption in the camps and at the same time, there is also widespread caste discrimination, and there are harassment cases, all these make it very difficult to make a relief camp function.

---

[9]A kind of lentil soup popularly consumed in India.

Ravages of Relief Activities **61**

Relief camps in Silchar thus were far from being the contemporary representations of Noah's Ark that Lal (2019) describes them to be. The context from which Lal was speaking about is one that is dominated by a strong history of left-wing egalitarian political activism, which has resulted in certain changes in the ways in which the bureaucracy works. The kind of unity and secularism that Lal (2019) and Ge (2019) describe has been possible in their contexts – Kerala and Tamil Nadu, respectively – because the governmental ideologies in place desired them to be so. In Assam, and especially in the Barak Valley, however, the days of egalitarian social justice-based politics have come to pass. The area remains plagued by a growth of a dominant far-right which has increased the propensity of Hindu–Muslim conflicts and caste-based violence in the region, which has also had its impact on the relief-camp dwellers, as two individuals from Muslim and Dalit communities, respectively, recounted:

> The government does not care about us anyway. They want us to die away. It is probably better that way as well. For most of them, we are just vote banks. They will come during elections and say that they will give us that, and that they will give us money, and jobs, and we will all vote for them. That is what we, the Muslims, have been reduced to.

> As Kaibartas, we are always the scum of the city. Nobody has ever treated us as equal members of the society. Most of our women have been treated like disposable entities during the floods, because most of them work as domestic workers. Some of their salaries have been cut, many of them have been forced to get back to work immediately after the floods. Same goes for men, not many employers have given them paid leaves. Most have cut money. Daily wage workers have been devastated. What will 3,800 rupees[10] do for them, their houses have been washed away.

The basic structure of a human society remains focused in and around the structure that the dominant community perpetrates in the society. In India, the dominant community has almost always been determined by the caste structure in place, with upper-caste males being the most dominant parts of the community (Chakravarti, 1993). They control the material resource distribution and the politics associated with the same and as such do not always look kindly upon the marginalised populace within an urban space. The people residing in relief camps

---

[10]The Assam government decided to pay INR 3,800 to each family staying at a relief camp. For more details, see https://economictimes.indiatimes.com/news/india/assam-floods-govt-to-pay-rs-3800-to-each-family-staying-in-relief-camps-other-shelters/articleshow/92578960.cms?from=mdr (accessed: 15.05.2024).

62   'Natural' Disasters and Everyday Lives

often get labelled as 'dirty', 'uncivilised', and a scar on the neighbourhood as can be seen in the following statement made by a resident of Silchar:[11]

> These people make the area very dirty; they come in hordes and do not live how human beings should live in these places. The entire place now stinks of their stool and urine. It has become so dirty. In a few months, there is Durga Puja here, how the hell are we supposed to worship there? These people are doing all sorts of activities there including using that as a toilet.

The way in which the entire labelling process functions provides one with a glimpse of the necro politics that has become mainstream under contemporary capitalism. Most of the powers associated with labelling have been reserved by the dominant community in place. Techniques of labelling have an impact upon the process in which communities and groups live their lives and form social groups or clusters – members of which are likely to have some similarity based on certain social attributes while the differences are often ignored (Zerubavel, 1991, p. 17). Social and spatial practices are entwined with lived experiences and the idea that individuals form based on these experiences (Lefebvre, 1991a, p. 34). They are influenced by the kind of community that the urban space produces as they themselves influence the communities that they become a part of or are infused into. The construction of social reality by the individuals actually living it remains contingent upon the various facets of their everyday practices and sensibilities (Giddens, 1991).

The differences between the ways in which various individuals experience everyday life give rise to certain codes, forms of knowledge, and means of interpretive action (Suttles, 1973, p. 36). These codes are sustained within the community through layers of horizontal and vertical relationships, which, as Fennema (2004, p. 435) suggests, are present in a combined fashion across the society. A key part of these codes during a natural disaster is constituted by anxiety and fear, which are one of the major drivers of anti-social behaviour. They affect the ways in which one interprets and engages with the world, as well as become one of the key components of the psychological effects. The anxiety associated with the fear that one might lose access to the basic necessities of life leads to tendencies such as hoarding and black marketing, often at a personal level and sometimes at a social level. A resident informs:

> Water stocking is necessary. Many people do that, but the middle class mostly. During the floods, they came to know that stocking and saving is so necessary that you need to do that in order to feel safe. So, these days, everybody has some bottles and rice in stock with them during the rainy season. Earlier, we needed these

---

[11]These comments which target a specific caste or religion can result in legal cases against the interviewee. Hence, the name has been withdrawn.

Ravages of Relief Activities **63**

> Kaibarta people to bring us these things during the floods, these days most of us have that, so nobody needs them. Plus, they make the area very dirty, it smells bad and is chaotic. They should get the camp out of here. It is a good society, not a garbage bag.

The effects of these tendencies in the society are felt most acutely within the relief camps where a majority of the vulnerable populace live during a disaster. Relief camps, though essential, do not receive the support that they need from the local inhabitants of the area. In the absence of much support from the government, the relief-camp dwellers suffer from various forms of material and structural violence. Despite being protected by the state and the local municipality in place, the people in the relief camps continue to get harassed and looked down upon by the local populace. A key part of the harassment received by them is constituted by the caste and class contradictions that they face in their everyday lives. As a Dalit individual who had been living in a relief camp during the floods stated:

> Whom would we go to? The community leaders told us that they had been setting up a relief camp here, so we came here. After coming here, we realised that there is corruption here also. Here also we need to pay to stay. The guard is saying that we will need to pay him extra to get what he calls a family room. Otherwise, he will put up in a room with twenty people. Government said we can live here, but here we are being stocked like animals.

It is no surprise that corruption is widely visible and makes its presence felt during disasters as well in Assam (Bhattacharya, 2003). The term 'family room' here, for example, means that an entire room is provided to a particular individual and the people with the individual. Although one might state that these issues are simple results of corruption, there is a larger caste–class–gender matrix operating here. In most relief camps, these rooms are mostly occupied by people coming from the relatively upper echelons of the lower classes. Most of the people who occupy such rooms come from the dominant sub-castes, are politically connected, and have some form of a relative class privilege even within their own sub-castes. Such events lay bare the myriads of internal contradictions that characterise the city of Silchar.

Desperate Acts of the Desperate Classes

During any major disaster, people frequently engage in a lot of different activities borne out of desperation. These desperate acts are often focused on survival and are usually mediated through commodities. This is because under capitalism, commodities are not just usable entities but are infused with exchange value that creates the value form – something that it draws from the very nature of labour under capitalism (Dunayevskaya, 1958/2015). Because the commodity inherits the inherent contradictions within the nature of labour under capitalism, the

64 *'Natural' Disasters and Everyday Lives*

commodity, 'in embryo contains all the contradictions of capitalism' (Dunayevskaya, 1958/2015, p. 85). Commodities are the soul of the capitalist structure in place, which makes them a fetishised aspect of contemporary capitalism. Disaster capitalism benefits from the commodity fetishism that is rendered mainstream under the capitalist mode of production in the society. Commodity fetishism is an intrinsic characteristic of capitalism and, as such, dominates the social relationships that arise under capitalism, at the core of which lies:

> [The] commodity form, and the value-relation of the products of labour within which it appears, have absolutely no connection with the physical nature of the commodity and the material relations arising out of this … (the commodity) is nothing but the definite social relation between men themselves which assumes here, for them, the fantastic form of a relation between things. (Marx, 1867/1976, p. 165)

It is this commodity form which gives rise to disaster capitalism frameworks to emerge during disastrous times. The commodity form, and its domination within the society which gave legitimacy to the alternate kind of an economy that was set up within and through which these camps which get transformed – often overnight – into bustling centres of trade and commerce. Some of the people who possessed a surplus of essential items resorted to selling them to their neighbours, community people, friends, and the relief-camp dwellers at an increased price. The primary sellers in this context were mostly middle-class people who had the requisite financial capacity to mitigate the effects of the floods. The relief campers became the secondary sellers who catered to the broader market in place keeping the commodity-based social structure functioning in the society. A middle-class individual and a relief-camp dweller narrate:

> I had a few extra packs of biscuits and water bottles from a function that has been held at my place a few months ago. So, I sold them at a bit of a higher price. I sold them to a local vendor, I did not take money immediately from him, but asked him to pay me later once he has sold off everything.

> What could we do? No income and staying in the camp had made life difficult. We are entitled to receive a certain amount of rice and lentils, but on most days, we do not get that. Mostly we get much lesser than what we should get. Those who have friends among the officers get more. So, I took to selling things to make ends meet, because ultimately you have to feed your family.

These events aptly become an example of Mbembe's (2019) observation of a necro-political framework wherein the social solidarity concerning human beings has, under capitalism, degraded immensely leading the world down a path where the problems of others do not invoke compassion but rather a sense of self-preservation, in the case of the relief campers, and a feeling of respite based on the idea that one oneself is not affected by the same, in the case of the middle class who acted as the primary sellers. The anxieties caused by material losses and a fear of the unknown propel most human beings to engage in activities with which they would not have normally been engaged in. These acts of individuals constitute acts of desperation. However, these acts of desperation have a distinct class character and are not free from the vestiges of a capitalist and casteist society. As the two individuals above narrated, a few members of the middle class saw in the floods an opportunity to get rid of the excess objects that they had in their possession – often for a profit – while those from the working classes saw in the floods, a struggle for their survival for which they had to engage in many practises that did not conform to their traditional occupations. The floods for them paved the way for a new mode of labouring within which for a brief period of time, they could transform into the exploiters of the middle class.

Many daily-wage workers in the wake of the loss of their everyday income turned to practising some form of small businesses to keep themselves afloat. These small businesses usually conformed to the disaster capitalist framework that scholars such as Klein (2007) and Lowenstein (2015) had put forward, which is focused on the garnering of profits during a disaster by major corporate houses and neoliberal governments. However, these individuals are different from the mainstream disaster capitalists because they do not get entrenched within a cycle of profit accumulation. For them, the engagement with disaster capitalism is merely a means of survival.

Under 'normal' industrial capitalism, capitalism ensures that the large capitalists or the monopolies go onto either control or completely wipe out the small capitalists (Marx, 1844/1973). This makes the large capitalists garner enough human and financial capital to engage with disaster capitalism. During the 2022 floods, however, the large capitalists could not exercise a direct control over the society because the enabling conditions of the same were disrupted by the inundated state of the region. Most of the large capitalists in the region found their avenues for direct profit accumulation stunted by the devastation that the floods brought in. Most of the large businesspeople in the region – ones who had controlled the region's consumer economy for years – found themselves unable to satisfy consumer demands because their supply chain had been blocked. However, the absence of the large capitalists and the disruption of the general supply chain in the region prompted a scarcity in the region. People subsequently turned to fair price shops, which were also unable to cater to their demands because 'the entire governmental structure had collapsed under the pressure of the floods and food was often available only within the relief camps'.[12] Ultimately, thus, it was left to the hoarders and the black marketeers to profit from the same, many of whom came from humble backgrounds.

---

[12]Statement by a fair-price shop owner. Name withdrawn on request.

66 'Natural' Disasters and Everyday Lives

Because of the roads being blocked, and the private exchange-based economy unable to serve the population, the 2022 floods gave rise to these sellers who engaged with the crisis in a way that satisfied the demands of the consumers. The alternate economy set-up within such disaster-struck areas becomes the basis of newer forms of disaster capitalistic tendencies to crop up within the society. Relief camps become the epicentres of such new tendencies because relief camps harbour the most vulnerable populace in a flood-stricken area in the South of the Global South (SGS). Typically, relief camps are seen as being the state's planned response towards any disaster. They are seen to be the most important and easily visible parts of the state planning mechanisms in place, because of the strict guidelines that are in place regarding the setting up of relief camps (NDMA, 2019). However, that being said, planning is not something that is necessarily anti-capitalist in nature. As Peter Hudis argues:

> [Most of the dogmatic Marxists] continued to define capitalism as 'unplanned' and socialism as 'planned'. The result is that 'socialism' became defined as a more equitable or efficient way of organising exchange. Left totally out of the picture is that exchange relations in capitalism are already 'organised' – albeit in a totally irrational way – by the law of value, which compels human activity to correspond to the dictates of socially necessary labor time. (Hudis, 2021, 'The Alternative')

The development of capitalist exchange relations within the relief camps proves the vitality of Hudis' argument that the primary aspect of one's critique of disaster capitalism cannot simply be the overhauling of the market by the state because the state also with time succumbs to the same logic of surplus value extraction because it does not negate the dominant exchange relations in the society. Within such a structure, all urban spaces are manifestations of the social fabric including the 'lifestyles, technologies and aesthetic values' which exists amid the people who design and live in the city (Harvey, 2008, p. 23). Often, those who design the spaces and those who live in it are segregated among themselves through lines of class, caste, gender, and race. Women among them are one of the worst-affected sections of the populace because they often lack the structural and ideological prowess required to mediate through the many exclusionary measures that the urban space implements. The urban space is critical to one's understanding of contemporary capitalism because it is the most important part of the capitalist structuring and restructuring processes (Harvey, 2012). The city, under capitalism, becomes the site of the most explicitly visible sites of the contradictions that exist between the elite and the marginalised. Take for example, the question of documents in the Barak Valley after the citizenship debates in the region. The citizenship acts-related amendments that were once proposed to be brought forth by the Government of India had made it necessary for many of the flood victims to save their documents first – many of which had been either lost or greatly damaged during the floods as Fig. 15 shows – rather than their other belongings because those documents are directly related to their livelihoods:

Fig. 15. Documents Destroyed By Floods in Silchar. *Photo credit*: Author.

>If I do not have the documents, they will cancel my voter and national identity cards. Already my national identity card is under processing after the whole citizenship amendment process was conducted. This time I had to leave behind a lot of belongings but at least I could save the documents. This document is essential for me to get a job and all the relief materials since the camp officers make it a point to see these documents.

The question of livelihoods that continues to play a role again brings forward the point that the major contradiction that operates within all capitalist societies is the contradiction between labour and capital, which also makes its presence felt during disasters, especially in countries such as India and more so within its highly underdeveloped and disturbed regions. The contradiction between capital and labour becomes the basis for the formation of class relations, and class relations – during a disaster – dictate whether one would become a hoarder or a petty seller of hoarded items. Both these processes, however, become everyday processes because it is the common people who engage in them. And anything that occurs

68 'Natural' Disasters and Everyday Lives

on an everyday basis under capitalism becomes a part of the mundane network of capitalist exploitation, because contemporary capitalism works in such a way that it engulfs the entire social existence of individuals. Negri (1982/1988, 2017) argues that such a state of the society is a manifestation of the post-Fordist society where cognition and immaterial labour have become a part of the capitalist totality in place mainly fuelled by an extension of the site of production.[13]

Floods and other such 'natural' disasters that take place due to the intervention of naturally occurring processes also become a part of this process. Different disasters play a different role in the process. Earthquakes are often easy to respond to because these are mostly once-in-a-lifetime events. Floods in Assam, however, are not. Assam and all its regions frequently get affected by floods, which has a great effect on the state's economic performance because of the state's overarching reliance on the agrarian economy. In the Barak Valley as well, floods are a regular feature which confronts the everyday life of marginalised individuals. It often becomes difficult for capitalistically oriented governments to respond to the effects of the regularity with which floods occur because such structures remain aloof from the everyday life of the marginalised populace. This was most beautifully explained by a woman during the floods, who had to shift to a relief camp on 22 June because the water supply of her house had been completely disrupted by the floods:

> What does the government know? They do not know whether we are living or dying. Most of them do not even consider us human beings. They just want votes. That is why the area is like this. All these politicians think we are just some pieces of entity that can cast votes for them.

The usage of 'natural' disasters for political gains is nothing new. In Silchar, in almost all the relief camps set across the town, there were huge hoardings of the ruling political formations assuring the victims that the politicians are with them. Most of these events tend to recreate the floods in the minds of the marginalised as an event or, in other words, a spectacle. In doing so, it reduces the entire process to an abstract universalist contradiction with minimum remorse for the suffering that the victims have to go through on an everyday basis in the camps.

---

[13]The Italian workerist thinker and activist, Mario Tronti, had realised this point back in the 1960s and had stated,

> [T]he relation between capitalist production and bourgeois society, between factory and society, between society and State achieves, to an ever-greater degree a more organic relation. At the highest level of capitalist development, the social relation is transformed into a moment of the relation of production, the whole of society is turned into an articulation of production, that is, the whole of society lives as a function of the factory and the factory extends its exclusive domination to the whole of society. (Tronti, 1962, para 17)

Guy Debord (1968) had argued that the basic character of the capitalist society in contemporary times is dictated by its fetish towards grand spectacles, whereby the process of othering becomes the fulcrum around which a society is constituted. The process of othering constitutes a critical part of the social factory because it creates the conditions under which capitalism destroys the solidarity mechanisms in place. For a region like Assam, this process gets easier because the region itself remains plagued by religious fundamentalism owing to the recent upsurge of religious conflict in the region (Bhattacharjee, 2021). The 2022 flood, amid such a political climate, became a spectacle for the broader Assamese society and political structure, which unlike the floods of the Brahmaputra Valley did not warrant national attention or hashtag activism. A Bengali mid-20s young person from Guwahati states:

> It is very difficult and weird to see that people do not pay any attention to the Barak Valley floods, that too in Silchar, the second largest town in the state. I know because my family is stuck there, and I have had to take help from friends to get them food and water. There is no question of travelling also as all the roads and railway tracks are blocked because of the rains.

The fetish with mega-events and spectacles is a result of the ways in which political communication has been implemented in contemporary India. The same is the case with regions such as the Barak Valley as can be witnessed in the case of the chief minister (CM). After numerous days of flood water rampaging through the region, the CM – Himanta Biswa Sarma – visited the region and gave an astounding speech to news channels which reverberated across the many Assamese television channels. However, the major focus of the speech was not on the scientific cause of the floods, or on the mitigation of the effects of the floods, but rather on the idea of how a certain section of the town's populace had been responsible for the same. Sarma squarely described the floods as a man-made disaster, one which had been brought into existence because of the activities of a few people who broke the dike at Bethukandi. Most of the mainstream media outlets subsequently focused on this statement from Sarma and enabled the creation of a Islamophobic discourse centred on the floods that reduced the floods to an act done by a few people from a particular community (Mehta, 2022) which contributed to the generation of an abstract notion surrounding the floods that does little to counter the concrete problems of development and governmental negligence that created the floods in the first place.

The developmental negligence was manifested in the stagnant conditions of life and sustenance that characterise the Muslims and Dalits of the region, who live in most of the low-lying areas of the region. Contemporary capitalism focuses on operating within the realm of the abstract contradictions without providing much attention towards the concrete issues that individuals face. When it does emphasise concrete contradictions, it is often only for garnering further exploitation through the more universal abstract contradictions. The focus on abstract contradictions allows capitalist political formations to steer clear from the internal concrete contradictions that form the core of the anti-capitalist struggles

70 'Natural' Disasters and Everyday Lives

(Anderson, 1995). The lack of everyday essentials produces the 'killer phase of capitalism' which significantly reduces the possibilities of a dignified human life and diminishes the chances of survival for the marginalised populace during disasters, converting them into victims and:

> The victim of a disaster is … unavoidably marked. As a victim of a disaster, is [one] part of its continuum? Is such an ill-fated individual also a sign, a fallen star in whose situation the suffering of others has been placed, as in the case of the scapegoat, or is that person, by coming upon the scene, a divine warning? Much of this turns on whether the victim emerges as a survivor. (Gordon & Gordon, 2007, p. 26)

The people at relief camps who had to stay for days outside their homes engage in various kinds of trade practices, which often go beyond the domains of legality. With many of them stripped off their regular modes of income, these individuals had to engage in other forms of exchange relations that often amounted to exploiting other flood victims. In doing so, these individuals do not only live their lives, but rather as Fine (2021) argues, they perform and present meaning, both to themselves and others. Generally, within a society, when an individual is in the presence of others, the immediate aim is to attain some level of acquaintance with the other individual's characteristic traits so as to create the possibility of getting a desired response by '[defining] the situation, enabling others to know in advance what [the individual] will expect of them and what they may expect of [the individual]' (Goffman, 1956, p. 1).

Within a disaster-struck economy, these expectations undergo a change in form and content, with most people desiring a radical change – a spectacular one – that changes the state of affairs at a rapid pace which suits the political manipulation of these disasters in a country like India where political communication has always focused on changes at a grand level. Guy Debord noted that the society under capitalism is a society that functions through the aegis of spectacles. It fails to solve the everyday contradictions that individuals face and instead implements its own version of solutions, which creates a break with the social reality that people experience. During the 2022 floods, the people of Silchar faced an uphill task of not only ensuring survival but also had to struggle to ensure that they were in a position to rebuild their lives once the floods were over. For a significant section of the populace, the floods had meant the complete demolition of many of their personal relationships in the city, especially those that were formed along broader communitarian ties. Much of these changes are related to the growth of neoliberalism as a social ideology. With neoliberalism, the changes that occur within the society go beyond the mere contours of economic relations. Neoliberalism, like capitalism, acts as a social ideology that introduces different kinds of changes within how individuals react to situations. It produces a social conditioning which disapproves solidarity and acts to create hyper-atomised social existence (Giroux, 2014/2020).

The hyper-atomised social existence makes them seek either for survival techniques or for avenues to maximise their income (or profits), rather than exhibit forms of solidarity. Within such a state, the basic expectation that people have from others is that of some kind of a benefit to emerge from social relationships. The expectations that individuals have from others in the community determine the extent to which they socialise within and through the broader community and social structures in place. In neighbourhoods with a high proportion of marginalised communities, it was the middle class which had been entrusted with the responsibility of distributing the relief materials. And the middle classes (and the middle castes) have a history of exploiting those lower than them within the socio-economic hierarchy, and the floods, yet again, provided a grim picture of the same. As a daily-wage woman construction worker displaced by the floods stated:

> They did not even allow us to enter their garage. We said that we will not do anything. There is no space in the relief camps, that is why I have had to come here. But then also, that person did not listen to us and gave a police complaint. What am I to do now? My house is gone, relief camps are full, and people with garages are not allowing us to enter.

However, there are exceptions to this general rule under capitalism as well, as two flood victims explained:

> The first people to come to our rescue were our neighbours, with whom to be honest, we speak very little. They came to our rescue, helped us get our furniture and electrical equipment upstairs on the terrace. The others came much later. It was not until three or four days that relief came to our doors. For three days, we had to survive just on what we were left with before the floods hit.
>
> It was I guess on the 23rd or 24th that I realised that I had not taken a bath for more than 4 days. With the heat, that was a serious health concern, more so because we were surrounded by dirty water. Then we went to search for a bathroom, but nobody was allowing us to use theirs. Then, it was near a relief camp that we found a house with a well, the owner of which had opened up the well for the general public to use.

For the people stated above, it was the neighbours – coming from a lower caste than them – and an unknown family who provided them with the necessary food and services. It is also important to mention that during the 2022 floods – when the water tanks of many were submerged under the floods – many people whose water connections were still working also opened up their private taps for others to take water from, as shown in Fig. 16. Such gestural interaction, though trivial in form, constitutes informal parts of a social contract in place (Vaneigem, 2012, p. 17), which is highly intersubjective and cultural in nature. This social contract does not only focus on the individual as a

Fig. 16.   Buckets and Tubs Line Up to Get Water From a Private Tap.
Photo credit: Author.

constituent element of social life but rather 'locates the social' within different structures of symbolic interactions or forms of communication which produce the 'social' (Reckwitz, 2002, p. 249). One of the major aspects of social life which relates to the issues raised by this intersubjective social contract is the question of routinisation of lives and the ways in which different social agents – individuals, groups, and structures within the society – interact with each other to produce a culture of their own.

Disruptions to Life and the Sustenance of Inequality

A natural disaster disrupts the normal flow of life in the society by creating shocks. A disaster wrecks the natural everyday lives of the people and the society and creates the provisions through which a form of unrealism establishes itself as the focal point in the society (Debord, 1968). The 'unreal' within contemporary capitalism is constructed out of the domination that capital and the markets associated with it exert over the common people, completely engulfing the effects that they have on their everyday life. Disaster capitalism continues to prosper because the contemporary capitalism creates highly unequal societies that tend to exploit vulnerable people. And the ways in which states respond to disasters are part of the methodological arsenal of the process of the imposition of neoliberalism in society (Klein, 2017). Disasters, especially under capitalism are not new, and will continue to occur as history progresses. In a highly dystopic sense, capitalism enjoys the prospect of disasters because with every disaster – be it economic, political, or ecological – capitalism is provided with newer opportunities for profit maximisation.

Kelman (2020) argues that one cannot merely categorise disasters into those few moments of human tragedies, but that 'The disaster is these long-term processes, over years and centuries, not the short-term events, over seconds (earthquakes), minutes/hours (tsunamis), and days (hurricanes). The process of unrolling disaster is based on the long-term choices of people' (Kelman, 2020, p. 14). They have been an integral part of the development of capitalist profit accumulation procedures because of their effects within the realm of labour markets, patterns of conquest, and consumption practices (Scheidel, 2018). Disaster capitalism has thrived in India – throughout Covid-19 and beyond – because it could work through the complex web of the state and the community in India (Deb Roy, 2024a). This has also been contingent upon the kind of developmental practices which have been implemented in India, which have created the necessary conditions for the upsurge of such tendencies. Achin Vanaik notes:

> From the fifties to the seventies, India had followed its own distinctive version of the import-substitution industrialisation model, more inward-oriented and state-regulated than elsewhere. The class character of the state was likewise sui generis – a dominant coalition comprising all sections of industrial capital, substantial land-owners, and senior bureaucrats, in which state functionaries operated as overall coordinators. In the eighties, a maturing bourgeoisie, more confident of handling external competition, and a burgeoning 'middle class' – actually an elite of mass proportions – hankering after higher levels of consumption, pushed for a cautious integration into global markets. (Vanaik, 2001, p. 44)

The failure of the developmental welfare state in India can be related to capitalism's relationship with the process of uneven development (Desai, 1975). Capitalist development is often highly uneven in nature designed to favour those who possess the requisite socio-economic and cultural resources to utilise the effects of the developmental trajectory to their own benefits (Harvey, 2005a; Wainwright, 2013). Uneven development is a result of the ways in which capitalism constantly impoverishes certain regions for achieving development in some select regions for some select individuals leading to the large-scale commodification of labour power in the latter regions (Harvey, 2006). The neoliberal reforms have ensured

74 'Natural' Disasters and Everyday Lives

that India continues to suffer from uneven development, a characteristic feature of the development of capitalism in the 1990s globally. The Public Sector Enterprises (PSEs) had been encouraged to function in India keeping in mind the role that they can play in eradicating the problems associated with issues such as pre-capitalist modes of exploitation, uneven development, and the like faced by people in India. The public sector in a country like India that is plagued by high poverty, low nutrition and education, and lower rates of employment generation avenues is critical to the human development of the nation-state and its citizens.

The general lack of autonomy – both financial and political – of the public sector has meant that the government has been virtually rendered powerless in the wake of major disasters. The devaluation of the public sector in contemporary India reflects the failures of the industrial and developmental policies and practices that had been institutionalised in India. These policies were initially developed to ensure an equitable development for the entire country, and the PSEs were supposed to play an important role in that regard. With the coming of neoliberal reforms, however, the PSEs found themselves to be at their wits end. Gupta's (1996) work chronicles the trajectory of neoliberal development in its early years where the companies owned by the prominent national bourgeoisie such as the Dalmia Industries and the Reliance Group were promoted through the careful political usage of acts such as the Sick Industrial Companies Act of 1985,[14] and institutions such as the Board for Industrial and Financial Reconstruction[15] – the intentions of which were to evaluate the performance of the PSEs but ended up being used to suppress the public sector (Gupta, 1996; Roychowdhury, 2003). The neoliberal reforms in India have been instituted in such a way that the public sector continues to function in a state of chronic powerlessness with highly reduced financial and social capabilities similar to how they had been prior to 1947 under the British (Makhija, 2006).

The uneven development that characterises India embeds within itself potentially politically explosive factors in certain contexts because '[most] of India's economic development in the past four decades has come from the southern and western states ... which have become hubs for manufacturing and high-tech industries [while] the heartland states ... have shown little sign of catching up' (*The Economist*, 2022, para 5). The distinctiveness of certain regions has always played a key role in the uneven development that many regions in India face, both economically and socially (Prasad, 1988). Even within the various metropolitan areas, cities such as New Delhi, Hyderabad, and Bengaluru have been characterised by socio-economic growth at a faster rate than other areas such as Kolkata, Chennai, or Guwahati. Within such circumstances, the fate that has been

---

[14]See https://legislative.gov.in/sites/default/files/A1986-1.pdf (accessed 23.03.2024).

[15]Founded in 1987, the 'Board for Industrial and Financial Reconstruction' (BEFR) was a development finance institution owned and directed by the Ministry of Finance, Government of India, and was a part of the Department of Financial Services of the Ministry of Finance. For more details, refer to https://indiankanoon.org/doc/148916823/ (accessed 23.03.2024).

Ravages of Relief Activities 75

awarded to regions such as the Barak Valley has been characterised by political negligence, social apathy, and notions of exclusion, as an experienced trade union leader from the valley states:

> The Barak Valley has always been disconnected from the broader Assam state. There have been repeated clashes between the Bengalis and Assamese on the question of language.[16] But of late, there has been a growth of Hindu-Muslim question and a constant demonisation of the valley.

While in a federal structure, the central government does have power to intervene in these issues if the state government remains inept at doing so. However, that kind of a proactive attitude is usually exhibited when there are different political formations at the centre and at the state level. As of 2024 April, both the centre and Assam are ruled by the far-right Bhartiya Janta Party (BJP), for whom the question of religious conflicts forms one of their fundamental agendas. The federal structure of the nation ensuring proportionate rights to the central and state governments was an integral part of the imagination of a free nation devised by the early Indian Republicans and Social Democrats such as Nehru and Ambedkar (Ram, 2014). One of the major questions that has haunted Indian federalism and the structure of Indian democracy itself is the question of uneven development (Desai, 2015).

Neoliberalism has resulted, as Harvey (2007) notes, in a situation where uneven development has been taken as a norm in the contemporary society. The results of the uneven development in India have been felt across the various regions of the country, especially within the north-east. The north-eastern region of India, comprising eight states, remains highly underdeveloped with respect to the broader country. The implications of underdevelopment become graver for regions which are seen as a misfit within the larger ideology or identity of the state to which it belongs to, especially after the linguistic reorganisation.[17] For a state like Assam *that is believed by many to have been formed for the Assamese*, the Bengali-dominated Barak valley represents a crisis of the ideology of the state. As a state, Assam is dominated by the mainstream Assamese-speaking populace, with very little attention being given to regions such

---

[16]For more information, see Deb (2019). Also see https://indianexpress.com/article/north-east-india/assam/not-assamese-not-bengali-there-are-bigger-issues-that-plague-assams-barak-valley-5197350/ and https://www.business-standard.com/elections/lok-sabha-election/citizenship-connectivity-unemployment-key-issues-in-assam-s-barak-valley-124042300328_1.html (accessed 01.05.2024).

[17]The linguistic reorganisation of Indian states took place in 1956, dividing states on the basis of the language that is mostly spoken within them. For more details, one can refer to the report by the Government of India available at: https://www.mha.gov.in/sites/default/files/State%20Reorganisation%20Commisison%20Report%20of%201955_270614.pdf (accessed 02.06.2024).

76 'Natural' Disasters and Everyday Lives

as Barak Valley. This has made the Barak Valley develop at a slower pace than Guwahati – both in terms of economic development and in terms of the human developmental practices.[18] The spaces which develop at a smaller pace exhibit a different model of communitarian practices than others.

In areas with high developmental index – economically – community-based practices can often be highly mechanical in nature, exhibiting higher tendencies of being engulfed by a profit matrix. Smaller towns, on the contrary, provide a much greater sense of place than the larger ones generating greater affective identification among the people (Belanche et al., 2021). Small towns are much more capable of developing a local sense of place than large metropolitan spaces because of the interactions that take place within these spaces within these societies are much more ingrained within traditional communitarian spaces, which needs to be recognised within management practices if they are to be developed effectively (Mattson, 1989). This also needs to consider that there is a growing pressure on the underdeveloped regions in terms of climate change because they frequently have to bear the impacts of the environmental problems often created by the more advanced regions (Narain, 2017; Singh, 2009). Growing urbanisation has accentuated the differences because of its role in economic development focused on the availability of cheap labour, structural violence, and exclusion.

The urban space in India is disproportionately tilted in favour of the elites and the urban middle class (Desai & Pillai, 1970/1990). The process of urbanisation in India has been fraught with tendencies of crisis, ranging from the existence of large-scale slums to the promotion of the most abject forms of caste and class segregation, even within the urban spaces. In the Barak Valley as well, such tendencies have become visible with slum dwellers being more affected by the floods in urban spaces, along with people from the hinterlands. The major cause for the same has been a constant weakening of community welfare measures which has contributed to a lowering of the autonomous powers being vested with the community which has in turn resulted in lesser community cohesion. This created the ideal ground for disaster capitalism to penetrate into the society through the usage of the avenues which are constructed by such 'shocks' – a term made popular by Naomi Klein's (2007) classic study of the disaster capitalism complex that talks about how global corporations benefit from disasters such as climatic events, wars, and the like.

Disasters have been known to have an impact on the structure of communities and states (McNeill, 1976). Under capitalism, disasters end up revealing the material and psychological aspects of human tragedies rather than concealing them, which is the norm under routine circumstances (Diaz-Qui ones, 2019). When disasters take place, there are certain events that occur at a rapid pace in

---

[18]See https://www.sentinelassam.com/topheadlines/why-is-there-a-disparity-in-daily-wage-of-te-workers-of-assam-607468. Also see http://cdedse.org/wp-content/uploads/2018/06/9A-Report-on-The-East-Bengali-Community-in-Barak-Valley-Southern-Assam.pdf (accessed 01.05.2024).

the society – both at the individual level and at the social level.[19] These changes often bring about great changes within a society. People who are part of a congenial community are often better equipped to deal with these changes, because they remain better protected by the communitarian structures in place (University of Sydney, 2022). Disaster capitalism frameworks disrupt such communitarian structures in place and tend to bring in a hyper-exploitative framework based on the furthering of corporate goals to use the collective trauma (Klein, 2007). The community is an important part of disaster-mitigation systems under these circumstances, especially with uncertain monsoons and climate changes (Nagendra & Mundoli, 2023). The community can improve the ontological security that individuals possess by improving the structures of solidarity in place that can counter the gradual unavailability of critical services during disasters caused due to rapid outsourcing, increased prices, and lack of responsible authorities, most of which are parts of a disaster capitalism framework.

A major aspect of the disaster capitalism complex is the privatisation and outsourcing of essential services amid the horrors that disasters cause (Klein, 2007) along with a complete marginalisation of the problems that marginalised people face during and after disasters. The roots of outsourcing amid disasters dates back to the 1980s and 1990s, when the privatisation agenda

> [...] had successfully sold off or outsourced the large, publicly owned companies in several sectors, from water and electricity to highway management and garbage collection .... By the late nineties, a powerful move was afoot to break the taboos protecting 'the core' from privatisation. It was, in many ways, merely a logical extension of the status quo. (Klein, 2007, p. 288)

Under neoliberalism, the processes through which privatisation works is often integrated intimately with the ideology of corporatism being elevated from being an economic one to a social one. The generation of corporatism as a social ideology always takes place through a combination of policy decisions that are sourced through a mixture of internal and external financial and social dynamics in place. Social relations are often managed through a manipulation of the local social

---

[19]One of the instances that Klein (2007) narrates is related to the transactions between the Federal Emergency Management Agency (FEMA) and the State of Louisiana in the United States of America during hurricane Katrina. When the state requested funds as a means to develop a contingency plan for a hurricane-inflicted catastrophe, the funds were denied. However, in the same year, the FEMA was found to have awarded USD 500,000 to the Innovative Emergency Management to come up with a disaster management plan (Klein, 2007, p. 409). In case of Covid-19 in India, the trajectory is similar. India would always be at risk of being affected more by deadly epidemics and pandemics because of the ways in which urban infrastructure has been structured in India. This has resulted in extremely congested living conditions with little attention being given to aspects of public health (Deb Roy, 2024a).

78 'Natural' Disasters and Everyday Lives

structures rather than the market directly, and during times of disasters, these social structures often transform themselves into exploitative ones. The unique social conditions produced by such local exploitative structures ensure that the community gets ruptured from within, a phenomenon that has wide-scale implications for the society as a whole. It is also necessary to note that the conditions for this rupture, however, are produced through years of capitalist exploitation of communities and marginalised individuals.

Chapter 3

Disaster Capitalism and the Omnipresent Market

Floods in the Barak Valley have become – for a majority of the people – an annual ritual that comes in every year, wreaks havoc on their lives and livelihoods, and leaves them completely bereft of any support thereafter. Under contemporary capitalism, the most affected by any natural disaster remain the highly marginalised and underprivileged section of the population. A key part of the marginality that gets exposed during any natural disaster in urban spaces in the Global South is manifested through the conditions of life faced by people from the marginalised communities, which constitute the bulk of the urban poor. Urban marginality under contemporary capitalism creates the grounds for the domination of the market over all the different forces of social cohesion. It has become a space where marginalised individuals, more so if they are women, are constantly excluded from rights of equal accessibility and participation (Kern, 2020; Massey, 1994). The urban form under neoliberal capitalism is one that is dominated by an exchange value-based capitalist planning, which makes it extremely difficult for envisioning social justice-based organisation and political forms of social organisation (Purcell & Tyman, 2015; Swyngedouw & Heynen, 2003). These changes within the urban form make it extremely difficult for individuals and communities to move beyond the profit-based rationality that neoliberalism produces and sustains in the society.

Henri Lefebvre (1975/2020) argued that industrialisation and urbanisation are two of the most important drivers of contemporary capitalist reality and the reproduction of the capitalist relations of production. Contemporary urban spaces, dominated by the growth of cities, have been a space where profits and growth have been privileged above everything else (Rodgers, 2009). Such a simplistic and uncritical focus on the creation of wealth has created urban spaces that are highly segmentary and structurally violent in form. Within such urban spaces, the community that one is a part of has become a critical part of one's subjectivity within the urban space. Communities can play an important role in ensuring care to the most vulnerable (Means & Smith, 1994). And municipalities are a

---

'Natural' Disasters and Everyday Lives:
Floods, Climate Justice and Marginalisation in India, 79–102
Copyright © 2024 by Suddhabrata Deb Roy
Published under exclusive licence by Emerald Publishing Limited
doi:10.1108/978-1-83797-853-320241004

80 'Natural' Disasters and Everyday Lives

crucial part of the community in contemporary times. However, most municipalities or the Urban Local Bodies (ULBs) are controlled – formally or informally – by the dominant community in place, which makes them inefficient organs of ushering in participatory democracy in the society.

The ULBs because of their inherent reformist nature under capitalism often remain inept at grasping the nature of the changes that disasters bring in within the society. The inability of the ULBs has a grave consequence for the people because in countries of the Global South, the most effective response to disasters is not always mandated by the state but rather by the non-state actors that come to dominate the immediate community that the disaster impacts. These responses are a combination of the contradictions that the contemporary model of capitalist development faces regarding the struggle that ensues between community capitalists, i.e. the small businesses, and large businesses, i.e. Big Capital. At the local level in India, the distinctions between the state and the market are often blurred (Harriss-White, 2006). Community capitalists who emerge from within the ranks of the marginalised communities enjoy a far greater privilege in terms of access to the immediate market during disasters than the large ones. They remain much more ingrained in the social fabric than Big Capital thus being at an advantageous position to exploit the crisis than Big Capital. The kind of disaster capitalism that these individuals practice is different from mainstream disaster capitalism and is related to the ideas surrounding social individuality, moral structures, and identity, both intimately related to the class, gender, race, caste, and the like.

This chapter speaks in terms of the access to resources, placemaking, and belongingness and how these aspects come to play a role during times of disasters. This chapter engages with the processes affecting the creation of a petty-disaster economy in regions, where groups of 'small entrepreneurs' and/or 'black marketeers' crop up within the society. The chapter highlights that while under normal circumstances, these individuals' identities as workers dominate over their subjective selves, during times of crisis, their alter egos as consumers often come to dictate their subjective selves. Drawing from the broader theory of disaster capitalism, this chapter modifies it in relation to the social reality that marginalised people encounter in their everyday lives in the highly underdeveloped regions of the Global South.

A Society of the Market

The lifeforce of a capitalist system is the market that it creates, which is dictated by the law of value. The natural law of capitalism ensures the creation of monopolies, where the larger capitalist slowly *but steadily* attacks the smaller capitalists until it completely engulfs the latter creating a monopoly. Jawaharlal Nehru, independent India's first Prime Minister and a Fabian socialist himself, acknowledged this tendency of global capital and, to counter the same, initiated the statist planning structures in India. In the early years of Indian independence, state-based planning had been popularly accepted to be the most useful method in which the Indian society can be developed. Planning had become central to the Indian democracy. The failure of the emergence of a left-wing alternative to disaster management principles in India also owes itself to the kind of fetish that most left-wing or social justice-based political organisations in India have had towards the state, which has made them reliant on the planning-based interventions that the state makes during times of disasters. One can here, in the context of the book, take the example of the Fatak Bajar,[1] the primary commercial centre of the town, which was the last place to close down during the floods. The market was so important to the urban space that it continued to function even when 70% of the city was under water.[2]

Capitalism, and especially neoliberal capitalism, contrastingly, bases itself upon the generation of chaos which provides an ideal ground for capital to thrive and prosper. Capitalism has often relied on the creation of chaotic conditions in the society for it to create the necessary conditions that it requires to produce the neoliberal modernity that it desires to render mainstream in the society. Raya Dunayevskaya argued that chaos is central to capitalist expropriation in the contemporary times but so is the idea of planning, especially planning done by those in power. The ideas concerning planning and chaos have been a central element of the debate between capitalism and socialism, with the former been associated with socialist or state-capitalist tendencies while the latter with capitalist systems per se. The analysis, however, should stand somewhere in between this duality. Dunayevskaya noted:

> Yes, planning is essential to capitalism and has always characterised the factory production and production relationship for it is the wherewithal of extraction of the greatest amount of surplus value. No, planning is not essential, chaos is, because while within production there resides the tendency to go outside the limits of production, class relations and existing values impose a limit on it, which expresses itself in the anarchy of the market. At the same time capitalism can never really plan because its law of motion is impelled by reproduction according to socially necessary labour time set by the world market, and thus even if all conditions are met as to plan in factory, external planning as to market, and labour paid at value, the incessant revolutions in production of necessity mean the 'development of productive forces of labour at the expense of the already created productive forces'. (Dunayevskaya, 1949, p. 9217)

Under capitalism, both planning and chaos go hand in hand, complementing each other. The market produces a certain amount of chaos in the society which is sought to be countered by the effects of the community which again under capitalism is supposed to bring in further alienation and estrangement. Under the welfare

---

[1]Fatak Bajar is the main commercial centre of Silchar.
[2]In 2024, the Fatak Bajar area did not face much water inflow, but the market did operate at lower levels, which caused a temporary price rise in the region.

82 *'Natural' Disasters and Everyday Lives*

state, the communities played a critical role in providing the people with the requisite support in situations and circumstances, especially under conditions where the state was unable to do so. During the floods of 2022, many people recounted their previous experiences of floods, and how the state and the government had helped them in navigating through them. One of them – living in his 50s – stated:

> In the floods of the 80s, there used to be so much support, not only from the government, but also from the community. Today except a few people, nobody wants to come forward. Some fear that they will get Covid, others think that helping others does not really matter anymore and that they are better off within their own homes.

However, as the previous chapters have noted, the growth of neoliberal capitalism has caused an erosion of the autonomy and welfarist activities of the community structures in place. In India, the community structure's engagement with factors such as caste and religion makes it a part of the exploitative structure of capitalism. Planning is critically important to such societies, especially in a society plagued with disasters. Long-term planning has quite a large corpus of socio-economic and political benefits, but at the same time, it also produces the necessary conditions required for capitalist systems of production such as state capitalism to sustain. For most of the underdeveloped regions, the state and its involvement in economic matters has often been seen as beneficial because it takes over the traditional role of the market (Haba, 1979; Mitra & Parakal, 1979). However, in underdeveloped societies, the overarching role of the state also runs the risk of a complete overwhelming of the people's movements and actually existing community-based egalitarian politics. The relationship that the state shares with the people in these regions is one that is dialectical in nature. It provides them with a certain level of empowerment and security but at the same time also becomes the primary force behind their exploitation.

The effects of the state acting as a monument impact the fault lines in the society. In the context of this book, one can argue that there are three monumental structures which have come into being, the state, the community, and the market. The state became the fulcrum around which people looked for hope during those desperate times. Societies such as Barak Valley which are plagued by long-term developmental lacunae benefit more from long-term state-based planning methods, even though it might be characterised by the same vanguardist forces that characterise full-fledged neoliberal capitalism. That being said, the marginalised people are the most important actors who are capable enough to resist against the onslaught of capitalism in the Anthropocene and show the path forward towards a social justice-based ecologically sustainable world. This is because of the ontological prominence that highly marginalised groups give to the idea of replacing the capitalist system with a more sustainable and just socio-economic and political structure that works for everybody (Federici & Richards, 2018). Such movements, 'their alternative, being creatively pursued by alliances among the exploited (with women at the fore), can be characterised as a global, horizontal, subsistence-oriented, decolonised, communing political economy, or what [one can] call "ecofeminist ecosocialism"' (Brownhill & Turner, 2019, p. 5).

Such solutions, however, are yet to be proposed in the Barak Valley, which remains dominated by political formations for whom the region and its vulnerable populace serve little more than cannon fodder within their larger electoral agenda. Political formations in Barak Valley remain restricted to a focus on winning votes, which makes even progress in matters such as disaster mitigation look like a form of stagnancy, as one can infer from Miliband's criticism of parliamentary politics (Miliband, 1964). Most of the political formations – including the Indian National Congress (INC) and the Bhartiya Janta Party (BJP) – have treated the marginalised populace of the valley as merely cannon fodder. This is because any change within the management of disasters in contemporary societies does not occur merely through political changes. However, unfortunately, that is the solution that most of the anti-capitalist forces provide to the marginalised, as a victim of the floods states:

> Most of them come and say that if the Congress comes, then it will be better. For so many years it was Congress only, but did anything change? No. The same cycle is on repeat. When the INC was there, we had floods, had to come to the same relief camp, now that the BJP is there, it is the same story. Then also, we did not have any proper sanitation or healthcare facilities here. It is the same now. Yes, some things are better, like we have better information and all, but the basic issues are still the same. No adequate food, not sufficient drinking water and no proper medical care.

Political change often does not result in major changes in the ways in which the marginalised populace is treated as most such changes do not radically alter the social structure. This is true even for political formations proclaiming to be socialist. State socialism had been a practising principle of India's governance during the tenure of Jawaharlal Nehru. Nehru's tenure as the prime minister of India had been characterised by the dual existence of state socialist policies and repression over the autonomy of the working class in the name of the national development (Sen, 1994). Similar issues can be found in modern-day China or in North Korea, where the exploitation of the marginalised has occurred simultaneously with the development of an authoritarian model of semi-socialistic economy (Fields, 2012). The existence of state monopolies aids in the escalation of such conditions because state monopolies enable the conditions suitable for the creation of a compliant workforce, which again results in a compliant voter base critical to a country like India where the poor are much more likely to vote than the rich (Price, 2013). It is for this reason that natural disasters have often become a tool for political campaigning in the region with even, the Home Minister, Amit Shah, stating that if the ruling BJP comes to power in a successive term, they will eradicate the flood problem in the state (Deka, 2022). The social reality constructed by annual floods in the region is such that such electoral promises do not make any difference to the marginalised because their lived experience makes

84 *'Natural' Disasters and Everyday Lives*

them completely impervious to the grand claims that political formations make. For example, as a flood victim stated during the 2024 floods:

> These are all false claims made by the government. They have done nothing. If they had done something, how does the dyke at Bethukandi get damaged in three consecutive years? Why does it not get repaired? Will it be a regular feature now that with every 2-3 days of rainfall, the people of Silchar will face a flood like situation?

Grand claims made by political formations regarding the floods have become a part and parcel of the political fabric of the state with floods becoming a regular feature in the manifestos of most parties. The lived experience that they experience within the urban space is not only a result of their objective circumstances but is also tied to the subjective consciousness that they come to develop in a country like India. Their subjective consciousness is formed because of the combination of the social practices that they come to experience and perform themselves, which as Lefebvre argues is related to the space that they occupy because space determines both the material and philosophical tendencies that dominates the society (Harvey, 2012; Lefebvre, 1982). The dominant material and philosophical tendencies determine the modes of social practices which in turn determine the dominant nature of social interactions. Exploitation only gets manifested when it becomes every day in nature characterising the interactions that individuals have among each other and with the social space in general (Rathod, 2023). The everyday life that people lead, in turn, is not dominated by spectacular events but rather by mundane activities. The development of capitalism changes the nature of many of the mundane activities that human beings engage, and the most critical aspect of any culture is found in the most mundane activities that individuals engage in (Bragg, 2017). As Henri Lefebvre had said:

> Everyday life ... defined by 'what is left over' after all distinct superior, specialised, structured activities have been singled out for analysis, must be defined as a totality. Considered in their specialisation and their technicality, superior activities leave a 'technical vacuum' between one another which is filled by everyday life. Everyday life is profoundly related to all activities, and encompasses them with all their differences and their conflicts; it is their meeting place, their bond, their common ground. (Lefebvre, 1991b, p. 97)

Mundane activities exhibit the most innate forms of social relationships that exist in a society. They often become the portals through which strangers interact which is a critical part of the urban spaces. Perhaps, the most interesting part of city spaces is that even though they are occupied mostly by strangers, the inhabitants still retain the capacity to live a civil life without necessarily 'hugging each other as they pass on the street, [be fully] happy with their lives, and [agreeing] on everything' (Monti, 2013, Sect.: 'The Leisure of the Theory Class'). There are certain boundary-making practices within the urban space that make it possible for the urban space to sustain itself as a cohesive space. Boundaries which individuals create aid them in keeping themselves secured psychologically. These, often porous, boundaries enable the individuals to engage effectively with their solitary actions, which can be defined as 'sequence of behaviour enacted by individuals with no input of interference by anyone else from one move in the sequence to the next' (Cohen, 2016, p. 5). Hyper-atomisation that neoliberalism creates tends to make these boundaries that individuals construct opaque in nature, creating more atomised forms of individuality within the society. The social practices that emanate from such atomised existence naturally focuses on individual gain, without giving much importance to the kind of social practices that they give rise to in the society. This process was evident in the manner in which people often resorted to hurting each other to get to relief materials in inundated areas where they were being dropped from helicopters – one of which is portrayed in Fig. 17 – by the National Disaster Response Force (NDRF). A resident of Bhagatpur stated:

> We got the news from some of our neighbours that they were going to drop relief materials, some biscuits and water from helicopter, and so most of us were on our toes trying to get the best. But some people wanted to get two packets, and they started pushing other people on rooftops. It was very dangerous. Imagine somebody falling from the terrace. It was very dangerous and inhuman.

Social practices under neoliberal capitalism develop through practices of atomisation that favour the creation of a neoliberal subject formed on the basis of an anti-solidarity and a disjunction between the society and the individual disrupting the relationship of symbiosis that the individual and society share as Marx (1973) had put it. Reckwitz (2002, p. 253) states:

> A specific social practice contains specific forms of knowledge. For practice theory, this knowledge is more complex than 'knowing that'. It embraces ways of understanding, knowing how, ways of wanting and of feeling that are linked to each other within a practice.

Practices help in demarcating lives of individuals in accordance with their convenience and schedules, which is also a means of demarcating the space around individuals along with the meanings these spaces hold. The partitioning of space also reflects the desires of the ways in which one wants to partition their social life (Zerubavel, 1991, p. 6), with specific sets of practices for public and private life that contributes to their psychological security. The psychological security of individuals is an essential part of their ability to experience the space and time, especially during a disaster such as flood. Floods with all its associated effects on relocation and physical mobility disrupt the stability in place (Lal, 2019). It disrupts the nature of the interactions that people have with other individuals and the society in general. The most critical role in this regard is played by the streets.

Fig. 17. A Helicopter Carrying Relief Materials in Operation During the 2022 Floods. *Photo credit*: Author.

Streets exhibit the highest amount of spatial access and often set 'the conditions for public and private action' (Fine, 2021, p. 99). These modes of action extend even to conflict resolution within these spaces, through the application of civic norms and routines (Monti, 1999, p. 205). However, during the floods, the streets had ceased to become a meeting and interaction point but had turned into a giant marketplace where different kinds of services and commodities were being put up for sale. The curious point, however, is that this marketplace was formed under the pretext of helping those in need. Because of the electricity being cut off, a large group of shopkeepers who had generators, brought out charging stations where they took to charging phones of customers for INR 10–INR 30 (0.10 GBP to 0.30 GBP app.) per hour. Such activities were complemented by a growing conglomeration of individuals in front of these shops, which allowed the city space to regain some of its lost essence amid the isolation that floods had caused. City spaces, as Sennett (1977, p. 39) argues, are spaces where the possibility of strangers meeting with each other increases significantly. Such possibilities provide every city with a distinct character of its own, making it a manifestation of a space defined by a 'cohabitation of stranger(s)' (Bauman, 2003, p. 5).[3]

Natural disasters make it almost compulsory for these 'strangers' to work together with each other. However, a working consensus is not easy to achieve because of the pre-existing internal contradictions that continue to function in the society and also because of the lack of common public spaces during the floods. The 2022 Silchar floods entrapped people within their homes, making them look for alternate modes of boundary-making processes, many of which led them to become more active perpetrators of the disaster capitalism framework. This is a manifestation of the understanding that urban spaces do not exist in isolation but are rather parts of a broad social structure of modernity (Giddens, 1991). If one defines cities to be constellations of strangers and their interactions, it is also customary to take note of the effects of 'high modernity', which places individuals within processes fraught with risks which previous generations of the subjects did not have to experience (Giddens, 1991). That often means most individuals often have to rely on higher powers or individuals to make sense of the situation that they have in front of themselves. However, such tactics do not always prove to be helpful in cases where the scale and impact of disasters exceed historical records. An old flood survivor aged in his mid-60s stated:

> We do not know what to say. Children had come to us regarding what they should do. We gave them some general advice but most of them know that we also have not seen this much water in our lives.

An uncritical reliance on higher powers makes them restrict their own subjective and objective experiences to ones that are akin to functioning within a confined space. Under contemporary capitalism in India, the government does not have the necessary freedom to practise policies that benefit the people because it remains entrapped within a global neoliberal framework (Ghosh, 1995). In India, the neoliberal policies practised by the state have made the cities a key part of the circuit of global production networks relegating its own importance to that of the market forces (Patnaik, 1984). In most cases, the dominance of the market has transformed the urban space into giant manifestations of the entrapments

---

[3]Bauman argues:

> It is common to define cities as places where strangers meet, remain in each other's proximity, and interact for a long time without stopping being strangers to each other. ... City dwellers are not necessarily smarter than the rest of humans but the density of space-occupation results in the concentration of needs. And so, questions are asked in the city that were not asked elsewhere, problems arise with which people had no occasion to cope under different circumstances. (Bauman, 2003, pp. 3–4)

88 'Natural' Disasters and Everyday Lives

that a marginalised individual faces – economically, socially, and politically under a market society. Cities often, as Pinnock (1989) argues, can look like garrisons under such circumstances because of the kind of planning associated with them which does not give due importance to the interests of the marginalised and instead seeks to entrap them within an alienating urban culture through capitalistic urban planning. The city of Silchar as well is no exception to this norm. The city, entrapped within various forms of developmental lacunae, provides little option to its vulnerable populace to escape the conditions of marginality, which get greatly aggravated during a natural disaster.

The Emergence of Petty-Disaster Capitalism

The theory of disaster capitalism had emerged mostly in societal contexts which have exhibited some sort of either a direct developmental trajectory infused with industrial capitalism – such as the United States – or through occurrences in places such as Afghanistan, which have been associated with or have been affected directly by the politics of the larger developed states. The larger developed state in the context of the latter acts as a monument, which determines the pathways for capitalists in the developing or underdeveloped areas. Monuments and symbols of authority are related to the networks of power and authority within the society and thus continue to serve as symbols of active and passive 'supervision', be it of authoritarian regimes or of democratic resistance. Buildings, which turn into monumental institutions, embody the history of the space they occupy within themselves. The space around such monuments can only be truly understood by replacing the silence of the buildings with the noise of the living beings affected by the monumentality of the structure (Harvey, 1979, pp. 380–381). Monuments such as the state and the community possess the power to determine and form rules around acceptability, respectability, and accountability, three vital criteria for the formation of any community (Monti, 2000).

With a state that actively acts at the behest of the market, the community cannot always be an optimistic aspect in India. In many cases, the community that people become a part of is in itself the major driver behind their exploitation. In the context of India, most communities get intertwined with the existing contradictions of caste and gender producing a harsh social reality for most of the marginalised. Floods in Assam have often made these issues visible to the wider populace. In the context of the present book and this chapter, one can take the example of the rise of the petty production-based units during the Silchar floods, something that even continued well after the floods were officially over. During the floods, as the previous section noted, there arose a string of sellers who sold essential commodities at heightened prices. They were affected by the broader privatisation of essential commodities and services, whereby commodities such as medicines had begun to be sold for exorbitant prices. A mother whose child was suffering from fever during the 2022 floods states:

> They only gave a few medicines for us in the relief camps. I cannot give that to my six month old child. All the pharmacies are also

> closed, and the hospitals are saying that their godowns are under water and that they have no medicines. The pharmacies which are open are charging high prices which I cannot pay, and the people who have access to government medicines are charging extra as well. It is very difficult to obtain medicines in this situation.

Petty-disaster capitalism can be defined as being a mode of disaster capitalism which acts as an aftereffect of the major effects that neoliberalism produces. Petty-disaster capitalism is a social complex constructed through and by the society and its relationship with the state. In major urban centres in India, these tendencies became visible during the Covid-19 pandemic which continue to reverberate among the citizens. While the mainstream focus remained on the larger urban centres, the smaller urban spaces have faced most of the hardships because these spaces remain bereft of the platform economy and technological advancements that contributed to a certain easing off of the issues during Covid-19 in most of the major urban centres. The aftermath of a disaster often brings forward newer avenues for small capitalists to profit from, as was the case of the small hotel owners in Uttarakhand in the aftermath of the collapse of an under construction tunnel (Baruah, 2023). These smaller capitalists are not directly a part of Big Capital but rather act on their behalf of smaller private capitalists. A society often becomes the adequate candidate for disaster capitalism to thrive in because it represents a combination of low economic and human developmental indices. The relocation of people during the floods robs them off the avenues for engaging effectively with their known environments, leaving them with no other option but to act in a manner that is focused on the extraction of money.

Capitalism, as Lebowitz (2020) argues, acts as an organic totality where different activities of the people act complementarily to each other producing the capitalist totality in place. The growth of capitalism as a social force becomes more pronounced under neoliberalism which produces a dehumanising form of life that tends to de-historicise human subjects and alienate them from the spaces that they occupy (Dardot & Laval, 2013). Neoliberalism's coming of age in India has been praised by many, such as Jalan (2021) for its continued undermining of the public sector and the state, because to them, centralised planning in India since 1947 had given rise to corruption as a socially acceptable aspect of the Indian socio-economic and political fabric of the nation. The gradual weakening of the public sector or the state's welfare mechanisms took place under the pretext that a distinction of their services and production mechanisms from the state and its associated political fabric would result in lesser political manoeuvring of the public sector enterprises.

However, such simplistic causal analysis fails to take into cognisance that the buyer–seller relationship that neoliberalism promotes does not attack corruption structurally but instead makes it starker in the society as it gains a socio-political legitimacy because it is primarily focused on the question of access that an individual within a position of power possesses – the numbers of which proliferate under neoliberalism in relation to those who do not, as access is made more difficult for the marginalised sections by the constant introduction of austerity

90 'Natural' Disasters and Everyday Lives

policies (Visvanathan, 2018). The austerity policies that have been unfurled in India have resulted in a gradual disruption of the welfarist character of the Indian state. Under normal circumstances, people somehow manage their living by engaging in low-waged activities, but during times of disaster, these activities or employment forms remain absent which makes them look for alternate modes of employment. Most of the people shifted to relief camps suffer from a complete break with their general life, which places them at a distance from the broader society in general, making them reflect upon their own conditions of life. As two relief-camp dwellers state:

> I am not saying that the government should provide for us every day at every point of our lives, but when these things happen, the government should come to our rescue, right? Otherwise, what is the point of voting? What is the point of being a citizen? We manage every day, but it is not so easy during floods or earthquakes.

> The relief camps do not have enough food or water or even good number of bathrooms. This college is supposed to house students, some hundreds maybe, and so they have like around 10-12 bathrooms, and that too only toilets with no bathrooms to bath. We bath in the open, like how we do back in our slum.

The relief camps that are set up by the state are usually characterised by the same set of contradictions that one finds in the mainstream society, as Chapter 2 has noted. The needs exhibited by the vulnerable and marginalised during a natural disaster speaks volumes about the responsibilities that the state has to consider while devising effective strategies to counter the effects of such calamities on the marginalised and the vulnerable. The growth of exchange relations – based on the idea of exploitation – does not only signify a change of human morality at an individual level but also speaks about the general financialisation of the society that neoliberal capitalism brings in, where every social relationship becomes a manifestation of and is evaluated on the basis of exchange relations. In the context of the 2022 floods, this can most aptly be witnessed in the case of the innumerable sellers of water bottles and mosquito repellents. The proliferation of sellers led to a heightened black marketing of these essential goods, with an intense competition between different sellers for profits – some engaged with this trend for profits, while others did for survival – which resulted in the conversion of the streets to a space dominated by commodities and disaster capitalism emphasising profits and survival of the fittest.

When competition is left unchecked, it is inevitable that many sellers will try to sell a similar commodity for higher profits. This will create the grounds for the emergence of newer forms of monopolies using various factors such as economic desires, social needs, and the state of crisis. Marx had noted, 'It is essential that the immovable monopoly turn into the mobile and restless monopoly, into competition; and the idle enjoyment of the products of other people's blood and

Disaster Capitalism and the Omnipresent Market   91

sweat turn into a bustling commerce in the same commodity' (Marx, 1844/1973, p. 267). Marx had noted:

> The same commodity is offered by various sellers. With goods of the same quality, the one who sells most cheaply is certain of driving the others out of the field and securing the greatest sale for himself. Thus, the sellers mutually contend among themselves for sales, for the market. Each of them desires to sell, to sell as much as possible and, if possible, to sell alone, to the exclusion of the other sellers. Hence, one sells cheaper than another. Consequently, competition takes place among the sellers, which depress the price of the commodities offered by them. (Marx, 1849/1977, p. 205)

During the 2022 floods, however, one needed to extend Marx's theory of cheapening prices with his arguments focused on the generation of monopoly capitalism, because there were two simultaneous processes happening. Those who had access to materials garnered more of the materials and caused an inflation of the prices often creating a localised monopoly in certain areas of the town. Petty-disaster capitalism emerges out of the competition that is left unchecked within disaster-struck societies. There were, during the 2022 floods, many petty sellers selling essential items such as mosquito repellents, candles, and water bottles. Each of them sold at a different price depending on the area that one served and the audience that one catered to. For those serving middle-class residential spaces, the prices of certain commodities – such as water bottles and biscuits – doubled up, while for some other commodities – such as mosquito-repellent coils and medicines – were being sold for slightly higher prices than usual. Many people facing this kind of a price rise demanded that the state or the government should intervene in price control and provide these essential commodities at their basic prices. The state or the community as a monument not only brings people together but also provides them with an aesthetical sense of belonging to the space they inhabit. Monuments or institutions play a pivotal role in the shaping of the community around the space. An oppressive monument would reveal the hidden cleavages in the society, while an egalitarian one would play a role in furthering social harmony. During disasters, such monuments further the already existing fault lines in the society. In the context of the present book, both the state and the community have become major articulations of the power that capital continues to hold among the marginalised, often leaving them with no other option but to engage in forms of petty-disaster capitalism.

Petty-disaster capitalism refers to a situation where the community itself provides the framework through which disaster capitalism prospers rather than being a hindrance to it. Mainstream disaster capitalism depends upon the state – working for the market – and the market, often acting autonomously. It bases itself upon the complete submission of the community to the demands and rules proposed by the market. The complete subsumption of the communitarian principles within the overarching structure of the state creates the grounds for the emergence of neoliberal capitalism. However, in the South of the Global South (SGS),

92 'Natural' Disasters and Everyday Lives

the state still continues to exert a considerable influence, often through the community affecting the everyday life of the individuals. The everyday life of individuals, as Lefebvre (2008, p. 2) understands it, 'refers to the set of everyday acts' which are interlinked and components of a collective whole – the social totality. This social totality is modified, sometimes organically and sometimes forcefully, by the changes occurring within everyday human lives which results in alterations in its constituent social practices. Most of the issues that affect the kind of urban experience that an urban space generates for its inhabitants, which remains bound by the capital–labour contradictions. As David Harvey had noted:

> The class character of capitalist society means the domination of labor by capital. Put more concretely, a class of capitalists is in command of the work process and organizes that process for the purposes of producing profit. The laborer, however, has command only over his or her labor power, which must be sold as a commodity on the market. The domination arises because the laborer must yield the capitalist a profit (surplus value) in return for a living wage. All of this is extremely simplistic, of course, and actual class relations (and relations between factions of classes) within an actual system of production (comprising production, services, necessary costs of circulation, distribution, exchange, etc.) are highly complex. The essential Marxian insight, however, is that profit arises out of the domination of labor by capital and that the capitalists as a class must, if they are to reproduce themselves, continuously expand the basis for profit. (Harvey, 1985, p. 1)

The long quotation from David Harvey can be taken as an initiation point for the theory of petty-disaster capitalism. In expanding upon the theory of petty-disaster capitalism, Harvey's insights – which follows Marx's (1867/1976) own ideas on the role and importance of the commodity under capitalism – prove extremely helpful because it focuses on the basic contradiction that exists within an urban capitalist reality, the contradiction between capital and labour. The major effects of the 2022 floods were faced by those who were the victims of the capital–labour relationship in the society, i.e. the slum dwellers of Silchar. Silchar has three major slum pockets, Madhura Ghat, Itkhola Ghat, and Kalibari Chor (Dhar, 2015). These slums do not arise out of vacuum but are results of long-drawn processes of exclusion within urban spaces. It is not such that slums can only grow in unplanned urban spaces like they did during colonisation in India (Sen, 1970/1990), but they can also grow up in planned spaces (D'Souza, 1970/1990), making these informal settlements one of the many issues that India's urban governance has had to face in recent years. Slums are usually characterised by extremely low levels of sanitation, public healthcare, and lack of essential services such as electricity, water, and sewer pipes.

Most of those who engage in petty-disaster capitalism come from slums – a usual result of the sustained housing crisis (Pillai, 1970/1990) – which has become almost endemic to structures of contemporary capitalist urbanisation.

Disaster Capitalism and the Omnipresent Market **93**

Slums reflect the degrading state of urban infrastructure that most of the urban populace live through in their everyday lives with reduced physical security, worsening state of sanitation, and a constant socio-political vilification (Davis, 2006). The ways in which people live in such areas reflect the degrading conditions of life that most such spaces provide. A slum dweller who had been selling essential commodities upon being shifted to a relief camp states about the unchanging conditions of life:

> In the slum, there used to be no water, no health system in place. The same is the relief camp. For us, flood or no flood, it is the same life all over the place and time. We do not have anything here. In the slum, we did not have any health facilities in place, neither any provisioning service. The area also is highly congested, so during floods all the garbage from each other's homes flows through the streets and homes. It is very dirty on normal days, more so during floods. When we come to relief camps, we thought things would change, but then everything remains the same.

The kind of urban planning that capitalism promotes makes it almost impossible for poverty-stricken people to pay attention to factors such as sanitation and health when choosing where to reside. During disasters in areas such as Silchar, the same conditions get replicated in the relief camps (and in general) producing a certain stagnancy for the marginalised people that contributes to a de-humanised state of existence for them where disasters become normalised and undifferentiated from their general state of existence. Such conditions of life encourage these individuals to get engaged in petty-disaster capitalism whereby they use their life abilities to extract, as one of them describes it to be, 'the best of the chances that [they] have to survive'. Theoretically, these practices embody disaster capitalism but at the same time, unlike the mainstream disaster capitalists, these individuals engage in disaster-based exchange practices because of survival, not for profits. Despite the form of their engagement remains in synchronisation with disaster capitalism, the content is completely different. While mainstream disaster capitalists see natural disasters as an avenue for garnering profits (Lowenstein, 2015), these individuals see in natural disasters a situation where they have no option but to engage in forms of petty-disaster capitalism for their own survival.

Individuals who engage in such practices during any major disaster usually come from marginalised class, caste, or gender positions. They do not work for Big Capital and do not earn huge profits, but the profit that they earn allows them to survive the impacts of the disaster. The motive of survival makes these practices distinct from the ones that are practised by the large capitalists for whom profits are a mere means for further profit accumulation. For the marginalised of Silchar, the floods became an avenue through which they saw an opportunity to earn money, which was sought to replace the losses that had occurred to their traditional income-generating practices caused by the floods. Slum dwellers have come to characterise the 'Urban Poor' in India, who have proliferated in numbers because of the unplanned urbanisation that has characterised Indian cities after the neoliberal economic reforms creating newer forms of marginalisation based on class, caste, gender, the kind of work they do, and the kind of living spaces that they possess (Bose & Saxena, 2016). The unplanned urbanisation becomes one of the key reasons behind people having to live in areas which get inundated by flood waters which remain stagnant for a long period as shown in Fig. 18 – increasing health risks – and making it mandatory for people to walk through water, one of whom can be seen in Fig. 19 – increasing the probability of meeting with accidents – which during floods in Silchar is often dangerous because emergency health services are often unavailable.

The engagement of the marginalised populace in petty-disaster capitalist forms of trade is also dictated by the fact that most of these people do not have access to safe drinking water, proper living conditions, and hygienic waste disposal systems in the relief camps. Like all other urban spaces in India, Silchar too is dictated by a range of discriminatory practices, based on a diversity of factors – religion, caste, class, gender, ethnicity, race, and creed – the implications of which become grave as one passes through the travails of a natural disaster such as flood when even the basic services – such as drinking water and connectivity – become

Fig. 18.   Dirty Water in People's Homes During the Floods. *Photo credit*: Author.

Fig. 19. A Person Walking Through a Flood-Affected House to Get Water Bottles. *Photo credit*: Author.

very difficult to access, especially for the urban poor. Being from marginalised communities, many of them are not seen to be equal members of the community and as such cannot enjoy the socio-cultural benefits that a community might provide during times of disaster to its own members. For example, in certain areas, the Dalits in the area were purposefully denied of accessing the basic services until all the others had been served. These areas – such as the Ashram Road and Tarapur – saw an increase of caste-based structural violence that reflects the depressing state of affairs in the city, as a Dalit woman states:

> The major issue is with the harijan colonies, where poor people often live. We were not given any form of relief at the onset. Most of it came very late. When it came though, the amount was negligible, not even enough for one person. Plus, there were many slurs hurled at us, that we are dirty, we pollute the area, etc.

Their situation becomes particularly volatile during the floods or other forms of public health crisis, which undermine the regular socio-medical order of the day, changing the social perceptions regarding individual safety and security. For most of the marginalised people, 'natural' disasters represent nothing more than a disruption to their economic activities. As a woman who sold water bottles with her two sons – one aged 5 and the other aged 3 – states:

> This *bhela*[4] costs around rupees 500-1000 to build. At that point, I have no other option but to sell the commodities at heightened prices. I not only have to break even with the money I had spent in building the bhela but also to get sufficient money in order to feed my family. During the floods, everything had become so expensive, there was no food grains or water available in sufficient quantity. The floods swept

---

[4]A flat boat usually made of bamboo. It can generally carry 2–3 people with some commodities or luggage. It is a popular means of water transport in north-east India. In the Barak Valley, it is often referred to as a 'Bhura'.

96 *'Natural' Disasters and Everyday Lives*

off my only avenue to earn. I am okay with living in a camp, and doing all the stuff that I can. But at the end of the day, you need to earn, you need money to survive because money is all that matters. In this situation, there is no other option but to succumb. I took to selling water bottles, sold a 10 INR's (0.10 GBP app.) bottle for 50 INR (0.50 GBP app.). Did I do wrong? Absolutely not.

The act of selling water bottles for a heightened price does not come naturally, and under normal circumstances, they would not have engaged in that as well. The fact that she had been engaged in such an activity proves the desperation that she was in, as she herself says:

There was no other option. What option I had? My area has a few Brahmins and a few Kayashtas. They settled down all the Kaibartas in one particular area near the railway station and we had no source of income. Most of us work in construction or as domestic workers, the first work was completely halted, and we did not get paid the daily wages. The second one sustained but many of us got told by people that they will not pay us as we did not go to work for 10-15 days, as they were also short of money, god knows whether that is true or not.

As a theory, disaster capitalism concerns the whole of the society. But like capitalism in general, disaster capitalism too works differently depending on the specific circumstances that it finds itself in, more so in countries such as India. India represents a complex myriad of various different kinds of abstract and concrete contradictions that do not only challenge the established bases of such theories but also prove to be counterintuitive to their mainstream forms. For example, Naomi Klein (2018) argues that disaster capitalism can be countered with the active intervention of the state. However, in countries such as India, the characterisation of the state is very different from those in the West. The exceptionalism of capitalism in India has been such that unlike the West, capitalism did not work through a rejection of India's pre-capitalist traditions but rather embraced the same for extending its own domination (Desai, 1984). This allowed capitalism to use the traditional pre-capitalist value structures to further its own domination in the society.

The ways in which neoliberalism has affected the Indian society has resulted in a certain continuum to exist between the state and the market, with the former often compensating for the latter through the local structures of authority (Harriss-White, 2006). The nature of this continuum creates the ground for the creation of a substitutive state that does not intervene, regulate, or compete with the market (Berry, 2022) but only acts as a complementary force when required. The substitutive state in India has had an indelible impact upon the marginalised in urban spaces, who have been the worst affected of the crisis of living phenomenon that has been caused by years of capitalist development because they have been entrapped within feudal and pre-capitalist modes of exploitation in a capitalistically developing society (Desai, 1990, 1991). The complicated state of affairs that the Indian working class faces on an everyday basis made the Indian Marxist sociologist, Akshay Ramanlal Desai, to write that:

> Socially, India has been one of the most complex countries. It has a continuity of history and cultural heritage which extends back to millenniums. It is a country which has probably the largest number of pre-historic tries. It has been a stage on which an immense drama of contact and conflict, fusion and fission, of a number of ethnic shocks, primitive tribes belonging to all stages of development, civilised communities and religious and linguistic groups has been enacted with an intensity and duration, probably unparalleled in any country. (Desai, 1969, p. 105)

During disasters, such structures exert a considerable influence on the different kinds of intra-community and inter-community exchanges that take place. These are the structures and exchanges that allow petty-disaster capitalism to thrive, which in turn provide a certain legitimacy to the capitalistic way of life that contemporary capitalism provides. The community in these contexts becomes just another articulation of the market getting dictated by financial motives and commercialisation which produce an alienated sense of belonging for most individuals. In countries such as India, all grand theories have to be reanalysed because of the intimate dynamics associated with factors such as religion, caste, and race in India. Caste and religion become the basic fulcrums around which the Indian society is structured, and the same remains true for the Barak Valley as well. The most important sections of the marginalised populace in the Barak Valley are the Dalits and the Muslims. Dalits and other lower caste people remain confined to ghettoised communities in certain parts of the town, mainly the Ashram Road, Harijan Colony, Kalibari Chor, Malinibeel, Madhura Ghat, and certain other areas on the fringes of the town. The Muslims face a similar kind of a situation, with many of them living in certain specified parts of the town such as Aulia Bajar, Madhurband, Panchayat Road, and Fatak Bajar, among others.

The areas that Dalits and Muslims occupy remain grossly underdeveloped and highly congested even within the general underdevelopment that characterises the town of Silchar. The situation that they face constitutes a state of accumulation by segregation that focuses on heightened discrimination (Jamil, 2017). The modifications around theoretical models become necessary because regions such as the Barak Valley and its urban areas bring forward issues that fail to get adequately represented by major theoretical traditions. Disaster capitalism, for all practical purposes, merely means the generation of avenues for profits through and within disasters. Within a neoliberal society, which focuses more on the production of commodities and consumers, rather than on the production of human capital or making life better for the marginalised sections, the engagement of the marginalised in exploitative exchange relations proves to be a major hurdle in the egalitarian imagination of a future social order. People from the other backward class (OBC) and Dalit communities settled up in relief camps engage in disaster

98 'Natural' Disasters and Everyday Lives

capitalism out of a certain compulsion which makes the subjective aspect of their engagement with disaster capitalism very different from those disaster capitalists that mainstream Global North theorists speak about. The engagement with petty-disaster capitalism becomes their method of engaging with the urban space, a space that has 'treated [them] like they do not exist'.

Disaster Capitalism, Corruption, and Peri-Urban India

Disasters and capitalism have a very curious relationship. Disasters bring forward the most vicious aspects of capitalism in the society, exposing the exploitative nature of social relationships under capitalism. The major manifestations of such issues are the prevalence of the highly vulnerable and marginalised individuals who populate the slums and other informal settlements which remain the worst affected sites during any natural disaster. The inability exhibited by capitalist societies to mitigate the effects of disasters for the marginalised populace manifested itself in the form of deteriorated conditions of access to essential services and commodities during the 2022 floods in Silchar. Slums, as has already been stated, remain one of the major manifestations of the problems concerning accessibility with most of their inhabitants, living under extremely low sanitation and health standards, unable to access the basic services. The growth of such informal spaces contributes to deteriorating state of public health and sanitary conditions and reflects a failure of urban planning (Bisen, 2019; Deb Roy, 2024a). Slums and other forms of informal settlement are a result of a combination of capitalistically planned urbanisation and population explosion (United Nations, 2003). The United Nations stated that as of 2001, around 924 million people had been living in slums, which had increased significantly since the 1990s. It projected that by 2030, the global number of slum dwellers will increase to approximately 2 million if no action was taken. The pioneering Marxist sociologist, A. R. Desai, had brought out the abysmal conditions of slums in India back in 1991 itself, when he had stated:

> The path of development pursued by the ruling classes has generated a distorted economic development resulting in a concentration of assets in few hands on one side, and pauperisation of the majority on the other, driving them to the tensions of unemployment or underemployment, lumpen activities, begging, garbage collection, etc., thus creating diverse situations with respect to shelter, essential commodities, and services. (Desai, 1991, p. 175)

Such a mode of economic development also has had its impact on the capacities that a society has in reacting to different kinds of natural and man-made disasters. The Indian state today, simply put, does not have the necessary and requisite financial and social capacities to react to a disaster like that of a flood without forcing the affected people to engage in tendencies such as petty-disaster capitalism. Petty-disaster capitalism is much more individualised in form and content than mainstream disaster capitalism. While disaster capitalism mostly

Disaster Capitalism and the Omnipresent Market   **99**

operates on a grand scale, petty-disaster capitalism in the SGS operates through the microcosms of the society, neighbourhood, and communities. It is often a means through which marginalised individuals exercise their right to the city that they live in. Through the practice of petty-disaster capitalism at a local level, the marginalised individuals desire to initiate and exert a certain right to the urban space that they come to occupy. In doing so, they also take up some of the responsibilities concerning disaster mitigation in their own hands. These responsibilities include minimum requirement of food, shelter, medical care, sanitation, along with extra attention towards widows and orphans (National Disaster Management Authority (NDMA), 2019). The NDMA also mentions that additional relief materials should also be provided if necessary, and that all relief centres should provide 'Ex gratia assistance on account of loss of life as also assistance on account of damage to houses and for restoration of means of livelihood'. However, if one goes by the narratives of the relief centre dwellers, it will be quite easy to see that most of these directives have not been followed:

> There was no discussion of when we will get what had been promised to us. Additionally, there was a large scale shortage of essential things as well like drinking water, food items, and medicines. Most of them did not take the full count of what was happening inside the camps. The officers do not know the entire truth about corruption and all. The ones distributing the materials used to give them first to those whom they knew, all of this was going on in broad daylight. We often used to get only the leftovers, and imagine, I have two little kids.

The management of the relief camps continued to be an issue in 2024 as well, when similar issues were raised by relief-camp dwellers pertaining to unfair distribution of resources, lesser relief materials available to people, and administrative negligence. Two relief-camp dwellers state:

> They have given us half a kilo or one kilo of rice, a little dal and some 200 milliliters of cooking oil. How are we to survive a family of three with that much? The water also is an issue. Only a little is given in the tankers. Most of them say that we should get it from the school, but the school tanks are empty.

> After making many calls, the relief camp management came today, but they only gave relief materials to four families. There are 350 families living here. They gave them because they are the relatives of the local **BJP** people, others like us can simply die.

While the administration failed to get adequate relief camps up and running during the floods, the people in low-lying areas had no other options but to flock to the relief camps. Many of them upon coming to a relief camp had to realise that the camp that they had been staying did not have enough resources – the

Normal School relief camp had no electricity and the residents had to contribute INR 50 (0.50 GBP app.) each family to get electricity – or was not registered formally as a relief camp itself, such as the relief camp at Cachar College for around two days (as shown in Fig. 20), which did not receive relief materials until around two whole days had passed from the time that refugees had entered its premises.

For purposes of disaster mitigation, two things are crucial, public services and welfare mechanisms, both of which are under threat under neoliberal modes of governance. For most of the marginalised sections of the populace, public welfare measures often remain the only ways through which they can access the services and commodities required during any major disaster to survive. The development of capitalism ensures that the large capitalists or the monopolies go onto either control or completely wipe out the small capitalists (Marx, 1844/1973). Disaster capitalism also focuses, and in fact depends upon, the elimination of small capitalists. However, in a country like India, this is difficult to obtain because the Indian state gives sufficient powers to itself to restrict these advances of big capitalists, mostly through its implementation of federalism and welfarism. Based on such ideas regarding federalism and democracy, the Indian economy had been structured to be a mixed one, and it has:

> [...] remained so after liberalisation. For, there occurred little privatisation, while the vast public sector continued to exist. ... At that level of generality, then, not much appeared to have changed. Continuity with the past is therefore one aspect that is apparent in respect of the impact of liberalisation. (Nayar, 2020, p. 134)

Neoliberalism in India, however, has led to a constant demonisation of the public sector. Scholars such as Das (2002/2012) and Jalan (2005, 2019) have

Fig. 20.　A Belatedly Registered Relief Camp During the 2024 Floods.
Photo credit: Author.

written about the problems that the public sector apparently posed for the rapid economic development of the nation through the institutionalisation of the bureaucracy, which they mostly ascribe to be one of the basic effects of Nehruvian socialism on the Indian society. A key role in this regard is played by the structure of the Indian federalism and the uneven development that has come to characterise India (Desai, 2015).

Neoliberalism has resulted, as Harvey (2007) notes, in a situation where uneven development has been taken as a norm in the contemporary society, which has made the idea of the right to the city a greatly relevant one in the contemporary times. The concept of 'Right to City' – initiated by Henri Lefebvre (1996) and later expounded upon by the Marxist geographer David Harvey (2012) – is one of the most prominent concepts to have come up over the past few years within anti-capitalist, or more precisely Marxist Urban Studies Theory. It is a part of the paradigm of urban theories which propose to go beyond the mainstream urban studies theories which often take recourse to capitalist normalisation to justify the bourgeoning urban centres and the process of exclusion which accompany them. The processes of exclusion become explicit during a natural disaster because natural disasters provide a window to understanding the degrading right to the city that these people enjoy. The demise of pre-existing communitarian forms of solidarity results in a complete subjugation of human relations under contemporary capitalism. That had been the focus of the theory of spectacle as initiated by Guy Debord (1968). This becomes evident in cases of some natural disasters where the site of the disaster becomes a media spectacle with government officials, tourists, and relief-camp dwellers (Baruah, 2023) engaging with each other on questions of destruction, without perturbing the logic of capital that produces these disasters. The transformation of a disaster to a spectacle undermines the seriousness that is required to mitigate its effects, which is through rigorous people-centric planning and participatory democracy-based management.

Urban centres are an important aspect of the development of capitalist economies and social systems. Urban areas are a part and parcel of the ways in which societies develop under contemporary modes of governance which often advocate freer markets and further accumulation of wealth in the form of profits. Urban centres under capitalism often develop through the exclusion of the marginalised people who perform the most arduous tasks in the developmental process of the centres. David Harvey (2003) has argued that capitalist urban development takes place through a process of accumulation by dispossession, which under neoliberal capitalism is furthered by a range of policy measures focused on privatisation of essential services and austerity policies. Austerity measures, in this context, means, as Sengupta (2011) argues, that there is a decline in people's capacities to manage their personal finances and a growing inflation. In a flood-prone area like Silchar, the increasing trend has become that one needs to live with the floods, much like the case in the broader Assamese society, as Das (2019) points out. It is popularly believed that 'learning to live intelligently with flood is more important than trying to drive out flood altogether – because nature will not listen to human commands' (Bhattacharya, 2003, p. 3), especially in regions like Assam, where the flood season extends from May to August every year.

102 'Natural' Disasters and Everyday Lives

Cities become extremely important in times of natural disasters because cities represent the highest stages of capitalist planning. They represent the ideal extent of disaster mitigation that planned capitalism can produce in the society in terms of the risks that the vulnerable associate with their everyday lives. Urban planning under capitalism has mostly been a neoliberal issue, one that has been dictated by rapid privatisation. Natural disasters reveal the relevance of the concept of the *Right to City* and the ways in which it still holds relevance in the society under contemporary capitalism – where the social structure is more exclusionary with the urban spaces and structures being planned and constructed in ways which benefit global capital and the elites. Corporatism has emerged as a global force in this regard emphasising the ways in which people can adapt to crises without making many adjustments to their own profit-seeking models. The corporatisation of the society acts against progressive *commoning* – the shared ownership of the public spaces and social labour – critical to the formation of any kind of non-hierarchical organisational form (Hardt & Negri, 2009). The influence of corporatist individuality also makes it difficult to invoke any kind of unity among the people. The unity of the marginalised is a critical aspect of the resilience that communities can show towards natural disasters and their aftereffects. Societies, in order to work through the effects of disasters, need to work through the marginalities that capitalism creates within the society and have to evolve from being mere platforms of capitalist assertion to more accepting and inclusive forms of urbanisation. The critical part to realise here is that natural disasters, although in a very limited sense, provide the human societies with a rupturing possibility to do that.

Chapter 4

Disasters and Everyday Life on Trial

The relationship between climate change and human society is one that has always been determined by a range of contradictions. Despite all the havoc that a flood wrecks in, floods do often have some agricultural advantages as well because of their effects on the quality of the soil, but that being said, they also have grave impacts on the people's lifestyles, especially those from the marginalised sections who get most affected by them (Bora, 2003). The most critical implications of these impacts are felt within the everyday lives of the people. Disasters become a part of the global circuit of production under capitalism. With each natural disaster that is unleashed upon the world, capitalism finds a way to make itself stronger, penetrating into the very heart of the human society. They enhance the effects of the major contradictions that the human relationship with nature faces today which is dominated by rampant industrialisation, consumerisation, and urbanisation.

Social theory plays an important role in unearthing these intricacies of contemporary capitalism and the life that human beings build within it. This is extremely important under contemporary forms of capitalist social structuring where the propensity of natural disasters being looked upon as divine interruptions within the human society have increased manifold. Rivers have often been looked upon as divine entities with a soul of their own that are connected to the practice of various religious rituals in different parts of the country. In the Barak Valley, the river Barak does not usually have any such religious connotations associated with itself, but it does have very specific social and cultural values for the people of the valley, like all south Asian rivers, which have always been important cradles of civilisation in the region. As Barman writes:

> Since the beginning of settled human society, the rivers of south Asian countries have intrinsically connected to societal evolution, especially in the context of the history of India and its neighbouring countries. From the Brahmaputra in the east to the Indus I the west and from the Ganga-Jamuna in the north to the Krishna-Kaveri in the south, these rivers and their innumerable tributaries have been accepted as the heart of Indian civilisation and culture. (Barman, 2023, p. 1)

'Natural' Disasters and Everyday Lives:
Floods, Climate Justice and Marginalisation in India, 103–123
Copyright © 2024 by Suddhabrata Deb Roy
Published under exclusive licence by Emerald Publishing Limited
doi:10.1108/978-1-83797-853-320241005

104 'Natural' Disasters and Everyday Lives

Human society is at a critical stage in the history of Planet Earth. The impact that the contemporary society and its various modes of progress have had on the Earth have ushered in a new geological era, the Anthropocene. The creation of the Anthropocene has also coincided with a rapid rise of science and technology, many of which have come at the expense of ecological and environmental sustenance. The development that capitalism promises works for a marginal section of the population, while consistently making a majority of the population vulnerable to climate and natural disasters. In the context of the Barak Valley, and Assam in general, floods constitute a key part of the disasters that people become vulnerable to. Floods are caused by an entire array of reasons, most of which are related to a rapid mismanagement of wetlands, farmlands, river water, and urbanisation (Viju, 2019). The ecological changes that take place within a river system owing to floods, soil erosion, high sedimentation, and incessant human intervention have created changes within the areas dominated by the riverine systems in place (Barman, 2023). These include the creation of environmental refugees, many of which go ahead and get engaged in commercial activities (Bhattacharya, 2022). The slum dwellers of Silchar referred to in Chapters 2 and 3 prove to be a testimony to this.

Bora (2003) argues that because of the people being unable to resist the effects of the floods, they are often forced to live with floods, adjusting their lifestyles to the probability of floods. Floods, as Das and Mitra (2003) have highlighted, increase significant health hazards and a reduction of social welfare. Under neoliberalism, India has suffered from a growing hunger problem, weakening public distribution systems, and increasing problems pertaining to public health which has also had an impact on the general income level of the society (Deb Roy, 2024a; Sarker, 2014; Swaminathan, 2000). Contemporary capitalism, working with a view to dismantling the welfarist remnants of the yesteryears takes a variety of approaches, one of which is the promotion of non-governmental organisations (NGOs), which is addressed in the first section of this chapter. The second section taking cue from the first discusses the objectification of lives under capitalism and the role that neoliberal corporatism plays in the process. The third and final section draws from these two sections and takes to discussing the importance of routinisation and the implications on the same during a natural disaster.

The Emergence of NGOs during Disasters

Neoliberalism eradicates the distinctions between the state and other private enterprises by implementing strategies focused on a corporatist framework managing them akin to other private corporations and generating interpenetrative approaches (Harvey, 2006). While, under a welfare state, the Public Sector Enterprises (PSEs) analyse their customers as citizens, under neoliberalism, they have to analyse them as only customers because of the changes occurring within their internal philosophy under neoliberal capitalism need to focus more of profits than on social well-being. Under the conditions presented by global capitalism, new public management (NPM) techniques have been used by the neoliberal forces to use the industrial relations existent in the PSEs to its process of profit accumulation encouraging public–private partnerships and austerity policies caused a general financialisation of the public sector and reduced its financial and social privileges and autonomy. The loss of financial and social privileges enjoyed by the public sector has meant that it remains incapable of making inroads into vulnerable communities even during times of crisis. The lack of financial and political autonomy becomes visible when, despite the best of intentions, the people from the public sector fail to respond to the crisis, those which also affect their own families. A water resource worker stated:

> We want to help them, but we cannot. Half of the area is under water. Earlier, we could pump it out. The only way to do that now is if we could somehow redirect the water to somewhere else, but that requires investments and a lot of infrastructural development. The state government does not pay us enough to do that.

With the public sector incapable of helping the people, and the absence of a properly functioning private sector, the NGOs stepped into action. NGOs and other Civil Society Organisations (CSOs) have come to play in the country, especially when it comes to disaster management and disaster responses. CSOs during the Silchar floods played a crucial role in distributing relief materials and helping many of the impoverished populace in getting access to certain commodities such as drinking water, charging stations, and power banks. The role that NGOs play during any natural disaster is critical to their functioning as NGOs and apolitical organisations because their activities during a disaster provide them with the much-needed social acceptance and validation. Their major contribution remains the promotion of community resilience through their activities (Park & Yoon, 2022). The growth of NGOs within the environmental movement was a political manoeuvring of the global climate politics which made the movement in the Global South compliant to that of the Global North that was dominated by the elite regions and their politicians (Ramesh, 2019). During the Silchar floods, the municipality took the help of NGOs in helping the people trapped in their homes, including many critical rescue operations as well (District Disaster Management Authority Cachar (DDMAC), 2022a). The requests received on the helpline numbers – shown in Fig. 21[1] – provided by the municipality were also, at times, redirected to CSOs.

CSOs enable a gradual corporatisation of politics with grassroots activism being replaced by a more advocacy-led programme. During conversations with a senior NGO activist, who has worked extensively with numerous grassroots NGOs, it was pointed out to the author that most grassroots NGOs have been converted to mere advocacy programme-based ones which focus merely on garnering more funding or aid and do little else.[2] A key reason for the transformation

---

[1]Phone numbers have been erased for protecting the privacy of the users.
[2]The activist had been a founding member of a prominent woman's rights NGO in India but has since been at a distance from them. Name of the activist and the NGO

106 *'Natural' Disasters and Everyday Lives*

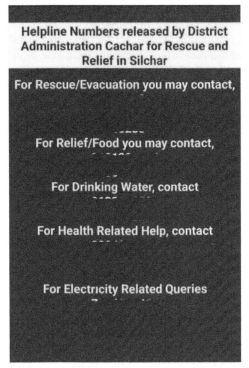

Fig. 21. Silchar Authority Helpline Posters Released During the 2022 Floods. *Photo credit*: Author.

is the growth of corporate interests and corporate funding within the NGOs, even in NGOs which had a significant amount of their membership drawn from left-wing organisations. NGOs often take the role that had been held previously by various actors involved within the Public Distribution System (PDS) structure in India. PDS is the basic framework that ensures that a particular society functions in accordance with the egalitarian values. The importance of PDS in India is immense as it helps the state in mitigating through the issues that the marginalised face in their everyday lives. However, the extent of the reach that the PDS has varies in accordance with the socio-political context within which it is implemented with regions having social justice-based formations at the helm of affairs performing better than others (Swaminathan, 2000). In a country such as India, an effective and well-managed PDS programme is essential because of the high rates of poverty that characterise the country that also translates into high rates of malnourishment and other health-related concerns. The pioneering Marxist sociologist from India, Akshay Ramanlal Desai, had written – way back in 1975 that, 'The Fundamental issue of the twentieth century is the battle against poverty, against disease, against illiteracy, against the economic misery

---

withheld upon request from the activist.

and the ruthless exploitation of the majority of population by a handful of capitalist owners in every capitalist country' (Desai, 1975, p. 104).

Desai's premonition has come true in the 21st century with contemporary India continuing to battle against the very issues that it had been battling against during the 1970s, albeit in a different form. The proliferation of CSOs in the country has resulted in a gradual weakening of the people's movements against such tendencies. NGOs and CSOs do not work for the people, as many would like to believe, but they rather work with a view to cause a gradual shift in the ways in which the society acts and reacts to capitalism (Hardt & Negri, 2000), which is evident and crucially important during times of crisis. Natural disasters have historically provided great opportunities for NGOs to make inroads into hostile political and economic social settings. During the Silchar floods, different kinds of NGOs also had taken up the responsibility that was once within the domain of the local municipalities. These activities include food/water provisioning, relief distribution, sanitation, and healthcare. The popularity of these NGOs during the floods had increased tremendously because of their apolitical nature. A male flood victim from the middle-class reminiscent about his preference for NGOs over and above their expectations from various political formations during disasters through these words:

> If we get help from the political parties or their people, they will expect us to vote for them, which again means that they will expect us to do certain things during the election season. That is very difficult for us. NGOs are better, they do expect some money, which we are happy to give. But then we do not have to do anything for them in return.

However, the critical point to note is the future political trajectory that the NGOs that the above respondents had spoken about have taken. Two out of the three NGOs have with time turned to their political counterparts adjusting and modifying their initial aims and objectives often with a view to becoming of the neoliberal social order. During natural disasters, NGOs play a crucial role, but most of the solutions that they provide remain bound by corporatist ideologies in place. Corporatism, in this context, means a method through which the relevance of political formations and the public sector is constantly attacked at the expense of the private sector and forms of apolitical or non-governmental (and non-radical) activism. In this regard, Schmitter (1974) proposed what is perhaps still the single most influential definition of corporatism:

> Corporatism can be defined as a system of interest representation in which the constituent units are organised into a limited number of singular, compulsory, non-competitive, hierarchically ordered and functionally differentiated categories recognised or licensed (if not created) by the state and granted a deliberate representational monopoly within their respective categories in exchange for observing certain controls on their selection of leaders and articulation of demands and supports. (Schmitter, 1974, p. 13)

108 'Natural' Disasters and Everyday Lives

Corporatism, as a concept, however, has not remained stagnant in the past decades. There have been changes within itself, which have brought forth newer models of corporatism based on the analysis of how corporatism can affect political structuring in the society by emphasising the weakened role of the state (Von Beyme, 1983). Out of the different definitions of corporatism that many authors have presented over the years, the current book will focus on the one that is presented by Hogan (1986). The reason behind the selection of this particular definition remains the clarity with which it puts forward the relationship between public and private interests as envisaged by corporatists under neoliberal capitalism. Hogan states:

> Organisationally, corporatism refers to a system that is founded on officially recognised economic or functional groups, including organised labor, business, and agriculture. In such a system, institutional regulating, coordinating, and planning mechanisms integrate these groups into an organic whole; elites in the private and public sectors collaborate to guarantee order, progress, and stability; and this collaboration creates a pattern of interpenetration and power sharing that makes it difficult to determine where one sector leaves off and the other begins. (Hogan, 1986, p. 363)

The growth of NGOs and other such depoliticised modes of resistance occurred along the aegis of the state moving out of the state of affairs in contemporary India is a part of the rise of corporatism as a social ideology amid disastrous events. It is important to note that disasters are events which cause massive historical and social impacts that continue to affect human societies for decades to follow. Disasters disrupt the natural course of humanity in a fashion in which the society could find little time to respond to the crisis. However, responses to all kinds of disasters – be they natural or man-made – harbour within themselves the fetish of outsourcing and extreme profit accumulation under neoliberal capitalism (Klein, 2007; Lowenstein, 2015). The characteristically distinct feature of petty disaster is that it makes these characteristics a part of the socialisation processes that people get engaged in and create in the society.

Wars, conflicts, and disasters have become usual elements of a capitalist world order. However, as against the mainstream world view, where even scientists such as Albert Einstein (1934/2012a, 1934/2012b, 1931/2012c) have repeatedly argued against wars and conflicts because of the impacts that they have on humanity as a whole, the Argentinian Trotskyist thinker Juan Posadas found hope in such apocalyptic events because he visualised them as a vehicle of social transformation. The primary drive behind Posadas's claims was his beliefs surrounding the existence of a superior group of beings – a hyper vanguard – who could come and rescue the human beings from the misery that capitalism exerts on the human society (Posadas, 1968, 1978). This becomes especially critical in countries such as India where disasters are frequently attributed to divine interventions into the lives of common people. The idea was similar to how many people believe in religious sermons and godmen, something which was evident even during the 2022 floods.

Disasters and Everyday Life on Trial **109**

Posadas' theories can get easily integrated within the corporatist goal of analysing the goal of achieving productivity, as Hogan (1986) notes, as a key part of social harmony. The focus on growth often results in the creation of a highly authoritarian and centralised politics which emphasise the growth of certain regions more than others. The effects that uneven development has on the extent of social cohesion can be diverse depending upon the particular context depending the kind of mainstream notions of individuality and the exchange relations that come to develop. Neoliberalism develops a very atomised definition of individuality that does not focus on the creation of wide-scale solidarity but rather works on the principle of generating far-reaching individualistic notions about meritocracy and achievements often at the cost of harming or sacrificing the well-being of others (Giroux, 2014/2020; Mbembe, 2019). Burns (1977) argues that a highly individualistic model of exchange relation-based society develops certain atomisation within individuals where they analyse their actions and relationships with regard to the society and other individuals or communities through a 'profit calculus' (p. 217). When an individual confronts a situation through which one cannot garner any kind of profit, it has – under neoliberalism – become a norm for the individual to refrain from such activities because of the atomisation associated with neoliberal construction of individuality.

Corporatism was conceived of as a method through which one could fuse the two seemingly contrasting modes of governance, namely the traditional welfarist models, and the paternalistic statist nature of governance (Hogan, 1986). It was supposed to be a path that moves beyond and negates the role of the state – both as a structure of authority and as a force of regulation – and accepts the role of the market as a force of social sustenance. The growth of corporatism at a global level has meant that regions with a relatively left-wing government have also had to resort to environmental decisions suitable for benefitting the status quo (Ge, 2019; Viju, 2019). This is because, as John Holloway (1995) argues, the source of developmental politics – money and power – still remains firmly within the grasps of the more developed countries and regions. This has a grave consequence for social cohesion in the underdeveloped regions. In such areas, as Amin et al. (2023) state, the flood-related relief mechanisms are based on the terms set by the developed countries that rarely counter the material realities faced by the people from the South of the Global South (SGS).

A key part of the problems with a homogenised tactic is the issue of underdevelopment, which makes it very difficult for people from these regions. Economic prosperity does not always result in greater social cohesion (McKenzie, 2011), but in spaces of the SGS, the lack of economic prosperity often results in people not having the luxury to work towards social cohesion. The lack of social cohesion, in turn, affects the behaviour of the people during natural disasters, which has become increasingly anti-solidarity and anti-communitarian in recent times, as Narain (2017) narrates. Social cohesion, according to Forrest and Kearns (2001), is primarily based on five inter-related dimensions of social existence: common values and civic culture, social order and social control, social solidarity and reductions in wealth disparities, social networks and social capital, and place attachment and identity. Some others such as Buck et al. (2002) argue that issues of social cohesion

110 'Natural' Disasters and Everyday Lives

are related to issues of social exclusion and social capital through three factors: social inequality, social connectedness, and social order. Most of these factors get altered with the coming of an event[3] with apocalyptic tendencies.

The 2022 floods in Silchar could not have been predicted, and hence, the reactions that the people showed towards its effects were also impromptu ones, but the construction of that promptness occurred through the many years of capitalist rationalisation of the society. The influence of the neoliberal reorganisation was felt during the 2024 floods, when most middle-class people had the time to prepare for the floods, but despite that reacted in the same exact way by succumbing to practices such as hoarding essential commodities, asking for disproportionate favours from their local community/political leaders and closing their gates for the displaced people. The submission of welfarist governments to the forces of corporate interests has made it necessary for one to rethink the basis of one's ideas on governments and governmentality. The growth of conglomerates in India has indicated the initiation of neoliberal capitalism in India, which has coincided with the growth of neo-fascism and other authoritarian forms of government (Teltumbde, 2018). The growth of the far-right politically means a diminishing autonomy of local political bodies, community organisations, free media, and democratic institutions. Most of these effects, as Mauro et al. (2023) argue, are complemented by an increasing attack on the environment because of the profits that the environment provides to capital and its associated markets. The most effective method that the people have against such tendencies, as Mauro et al. (2023) state, is through organisations that focus on the community as an unalienated and environmentally sustainable basis of organisational principles and do not rely only on electoral politics but rather advocate a complete change of the power dynamics. Rajeev Bhargava states that

> a government that satisfies the basic needs of everyone successfully is better than one that respects the political rights of its citizens but leaves their basic material needs unfulfilled. If so, impoverished people may prefer a minimally welfare-oriented authoritarian government to a democratic government that fails to redress poverty. (Bhargava, 2006, p. 320)

---

[3] The definition of 'event' in this context comes from Hannah Arendt, who defines 'events' to be

> [...] occurrences that interrupt routine processes and routine procedures; only in a world in which nothing of importance ever happens could the futurologists' dream, come true. Predictions of the future are never anything but projections of present automatic processes and procedures, that is, of occurrences that are likely to come to pass if men do not act and if nothing unexpected happens; every action, for better or worse, and every accident necessarily destroys the whole patterns in whose frame the prediction moves and where it finds its evidence. (Arendt, 1969, p. 7)

'Natural' disasters and the role that governments play under such circumstances prove to be one of the most important markers of the progress that a society has made. The people of Silchar's support for the far-right political formations cannot be seen as an isolated action that does not have any social implications, apart from the obviously political nature of the same. The kind of government that exists within a social setting plays a crucial role in the mitigation of disasters. A social justice-based governmental set-up prevents the commercialisation of a natural disaster by active involvement in the society, while a more liberal-oriented right-wing government sees it as an opportunity to extract profits and extend free-market propaganda. The NGOs can be a part of either one of the two above, but in Silchar, most of the NGOs succumbed to the latter, whereby their activities were instituted and formally recognised – even by the DDMAC – while the actions of radical organisations such as the trade unions and collectives were purposefully not acknowledged, as an activist explains:

> That NGOs take benefit of disasters is nothing new. But it has increased these days. In the absence of any organisation, certain trade unions took to the streets distributing relief materials and medicines, especially the medical sales and representative unions, but there is no mention of that anywhere, neither in the government outlets, nor in any of the media outlets. That is the domination that NGOs have today, that even government is functioning through the non-governmental entities.

a picture of being surrealistic, but their extraordinariness actually lies in their ordinariness or, in the words of Lefebvre (1991b, 2002), in their triviality. The problems with surrealist understandings are that they neglect the trivial which stems from their 'transcendental contempt for the real' (Lefebvre, 1991b, p. 29). The most extraordinary things related to human existence are to be found in the everyday life (Lefebvre, 1991b) completely subsumed by the forces which dominate the society (Debord, 1968). Everyday life is laden with the contradictions which have their own roles to play in the sustenance or the disruption of the existing social order. In other words, these contradictions and their dialectical relationship with the social space form the foundations of the social totality, be it in the hinterlands of India in the Global South or in the urban neighbourhoods of a prosperous country such as New Zealand in the Global North.

The Objectification of Lives

Capitalism works through an objectification of human lives. For capitalism, the human lives represent a reservoir of labour power, something that it continues to exploit till the reservoir gets completely exhausted (Marx & Engels, 1848/1976). Mostly, such exploitation occurs through the contours of class. But class under contemporary capitalism is just one aspect of the capitalist corpus of exploitation. There are multiple axes of discrimination that determine the extent to which one's life is valued under contemporary capitalism in India. These axes are based

112   'Natural' Disasters and Everyday Lives

on various different kinds of social and biological attributes such as gender, caste, ethnicity, race, and creed. The normalised construction of an 'Indian' is based on an individual being able to manifest certain qualities deemed essential for being considered to be a normal Indian citizen. As a Nepali person, whose home in the city's Jhalupara[4] had been washed away by the floods which had forced him to shift to his relative's place, stated:

> Nobody comes and visits us. Jhalupara is a long lost area. Most of the people think that you get only Momo there, and not any human being. This is such a massive issue, with our narrow roads and bad housing distribution. But nobody comes here. They think that we are all *Nepalis* living here, what help we will need. It is because we are poor and not traditionally Indianised enough in the general sense of the term that we face this. God knows what happens to the Muslims.

If one does not manifest the qualities mentioned above, an individual is often considered to be an inferior human being and is often left devoid of the benefits that a citizen would be deemed to receive from the state or the society during times of crisis. The implication of such discrimination becomes greatly overwhelming during periods of extended austerity-based politics that produces a certain stagnancy in the economic structure of the society (G, 2022). The growth of austerity politics is disproportionately faced by the marginalised because they cannot protect themselves from the economic implications of austerity measures resulting in higher mortality rates for them (Dorling, 2017). People from the marginalised communities have been stigmatised not only because of their socio-biological attributes but also because of the kind of work they do, which is often enforced upon them by their socio-biological attributes. For example, the institution of caste in India has made it difficult for Dalits to become integrated into the society as equals given the tremendous pressures they face to respect the upper castes and the socially enforced demonstration of that reverence (Guru, 2013). Despite performing some of the most important tasks within the society, Dalits have always been treated as inferior citizens in India. The treatment that is meted out to them is a manifestation of the structural inequalities that have plagued the Indian civilisation for centuries.

Capitalism's attack on ecology and its erosion of local histories affects the structures of social solidarity in place making it harder for people to organise themselves (Sethness & Clark, 2023), especially if they come from marginalised communities or are engaged in professions that are historically considered to be sub-human in nature. In the context of India, such professions include sanitation jobs, manual scavenging, and domestic work. For example,

---

[4]An area dominated by the Nepali population in the city. The area has congested housing and is located adjacent to the central jail and the Assam rifles cantonment in the city.

with very little political mobilisation around the feudal vulnerability and non-dignified nature of the professions such as sanitation, manual scavenging, and domestic work, they have become almost a perennial characteristic of India, something that even neoliberalism could not erase. Considering that the sanitation workforce in India is largely composed of Dalits, the discrimination and stigmatisation of these workers attacks both their social and economic well-being (Thorat & Thorat, 2022). These are the vulnerabilities that characterise modern India as a deeply divided society, which contains various kinds of cleavages characterised by different community claims to territory, resources, and superiority, with further divisions even within the communities on lines of class, gender, and other geopolitical attributes (Guelke, 2012; Lustick, 1979; Nordlinger, 1972).

The fault lines that exist in India make their presence felt during a major natural disaster where people from marginalised communities are often made to suffer from increased vulnerabilities. One such example might be the case of 'Flood Jihad' and the fears of the 'Kaibarta Takeover' which were being rendered mainstream in inundated Barak Valley. The chief minister of Assam, H. B. Sarma, stated that the floods of Barak Valley were not a natural disaster at all, but rather it was a result of the vandalising activities of the Muslims.[5] This statement had a wide-scale impact on the marginalities that the floods produced for the people. A BBC report stated:

> [For] a number of social media users, something more sinister was at play here. They claimed, without proof, that the floods were man-made, and that a group of Muslim men had deliberately inundated the neighbouring Hindu-majority city of Silchar, by damaging flood defences. Mr Laskar's arrest, along with that of three other Muslim men, triggered a barrage of social media posts accusing them of supposedly waging a 'flood jihad'. These posts were shared thousands of times, including by prominent influencers with verified accounts. The claims were then repeated by some local media outlets. But the gravity of his situation only dawned on him when, already in prison, Mr Laskar caught a mention of his name on television: a news channel accusing him of 'flood jihad'. [He said] 'I was afraid and could not sleep that night. The other inmates were talking about it. I thought I might get attacked'. (Arora & Silva, 2022)

The ways in which the Muslims are being marginalised in contemporary India get reflected in the treatment that is meted out to them that constructs a social situation where they are blamed for everything that does not go according to a plan.

---

[5]For a full report on the same, see https://www.facebook.com/newslivetvofficial/videos/750889339448488/?extid=CL-UNK-UNK-UNK-AN_GK0T-GK1C-GK2C&ref=sharing (accessed 20.04.2024).

114 'Natural' Disasters and Everyday Lives

The Muslim populace which was blamed for breaking the dike in 2022 – in the name of Flood Jihad – had to face the brunt of administrative negligence and corruption[6] which failed to repair the dike in time in 2024 as many Muslims were again blamed for the damaged condition of dike, as a Muslim resident stated in 2024:

> They are again saying that the Muslims have broken it. I mean what kind of treatment is this? The dike was to be repaired but no major work was done and now when it has again on the verge of collapse, they have started saying that Muslims have done flood jihad,. Do they even know the meaning, the true meaning of jihad?

Muslims in contemporary India continue to face extreme hardships with fewer recruitments in bureaucratic positions and often been dubbed as second-class citizens because of the domination that ethno-religious notions of Hindu supremacy have come to exercise politically in India since the early 1990s (Jaffrelot, 2011). The Islamophobia that is rendered mainstream in the society got further accentuated in India, with most of the media houses also actively partaking in the same by promoting the narratives about four Muslim men – Kabul Khan, Mithu Hussain Laskar, Nazir Hussain Laskar, and Ripon Khan – being solely responsible for the floods because of their alleged involvement in damaging a dike (Mehta, 2022). This narrative did not tone down even after a week after the floods, and Raja,[7] a fruit seller living in a relief camp because his house in Tarapur was still inundated, states:

> The Muslims have done this. They do not want our valley to prosper. They want low development. Because high development would rob them off the capacity of have too many children that they usually have. It is their fault. Himanta Sarma is absolutely right, we should have driven them out long ago.

The statement above by a Hindu man proves the extent to which the far-right ideology propagated by the *Bhartiya Janta Party* (BJP) and the *Rashtriya Swayamsevak Sangh* (RSS) has penetrated into the society. During the floods of 2022, the Muslims of the region faced grave threats over their lives and property. Many of the Muslims, as a daily-wage worker Abdul informs, 'had been forced to stay in their inundated homes or with other Muslim people in their homes, because they did not feel safe in the relief camps'. Such a situation faced by the minorities is nothing new in the context of a major disaster in India. The dire situations

---

[6]See the detailed report regarding the All India Trinamool Congress leader, Sushmita Dev's allegations: https://www.barakbulletin.com/en_US/trinamool-mp-sushmita-dev-alleges-bethukandi-sluice-gate-was-opened-for-ceremonial-closure-demands-investigation-into-funds-expenditure/ (accessed 05.06.2024).

[7]Name changed on request of the interviewee.

that Dalits and Muslims face under neoliberal capitalism in India, even when they form around 16 percent and 14 percent of the population, respectively, had accentuated even during the Covid-19 pandemic (Deb Roy, 2024a) and continues till date. A young Muslim person recounts:

> First there was Covid-19, and then there is this. Covid-19 they call it Corona Jihad, now there is floods, and they are calling it Flood Jihad. We are poor, that is why we get caught up in all the bad situations that there are. Most of us do not have much idea about what is happening around us. The dam at Bethukandi was broken, I admit it, but what would have those people done/Their homes were getting drowned every week.

It is important to mention that the areas surrounding Bethukandi dike are mostly populated by Muslims. Similarly, another dam which has caused major distress among people is the Shibbari dam, neighbouring the Kalibari Chor area of the town mainly populated by lower caste people, which continues to evade the attention of the authorities, as a local activist mentions:

> The problem is that the entire focus is on Bethukandi, mostly because after 2022 people have come to realise that the Bethukandi dike is critical for flood control in the region. But nobody is talking about the Shibbari dam

The Shibbari dam, it must be mentioned, is not a new issue but has been a long-standing one for many years but continues to evade the attention of politicians and administrators. An aged activist informs:

> The issue of the Shibbari Dam and the Kalibari Chor area is there since I was in school. We had also protested against it when the large flood of 1991 happened. But it continues to be raised only when floods are here. After that or before that nobody even bothers about those poor people who have to leave their homes every time a little bit of rainfall happens.[8]

The conditions faced by Muslims, Dalits, Tribals, and women aptly get reflected in the average life expectancies that one finds in these communities. According to Vyas et al. (2022), Dalits, Adivasis (Tribals), and Muslims in India are much more likely to die early. Adivasis, Dalits, and Muslims live for 4 years, 3 years, and 1 year less, respectively, than those from the upper-caste Hindu communities. Their analysis is based on the data collected from above 20 million people from

---

[8]The people from the region have to leave their homes almost every time whenever there is moderate rainfall. It is frequently taunted by the mainstream populace as being a dirty and underdeveloped area.

116 'Natural' Disasters and Everyday Lives

nine Indian states, which total to around half of India's total population. The same situation persists for Dalit, Adivasi, and Muslim women as well. Women in India cannot be a universal category because there are vast differences between how they are treated in different micro social contexts, in accordance with their caste and class (Rege, 2020). The life expectancy of Adivasi, Dalit, and Muslim women is 62.8 years, 63.3 years, and 65.7 years, respectively, while the upper-caste women live for around 66.5 years on an average.

Mortality rates in India are contingent upon the various social attributes and ascriptions that an individual possesses. Dalits are always more at risk of facing high mortality rates than upper-caste Hindus, which have remained the same over two decades after the neoliberal reforms (Gupta & Sudharsanan, 2022). After the 1991 economic reforms, which had resulted in the creation of a new middle class in India, especially in the urban areas, the situation of Dalits did not undergo any significant change in terms of micro or macro factors (Vidyarthee, 2014). Dalits and Muslims have found themselves to be increasingly vulnerable to natural disasters, not only because of the institutionalised structural inequalities but also because the response to most natural (or man-made) disasters do not pay adequate attention to their contemporary social and occupational marginalisation, as was also the case of the 2022 Barak Valley floods and which can be seen from the following statement from a Muslim resident:

> There are a couple of points which need to be mentioned. The first is that Muslims are in general poorer than Hindus.[9] Sadly, that is true. So, relief materials needed to be distributed proportionately. A Hindu person often knows political people, and they will also help the person because of their religion. That is not the case with Muslim people, we have to struggle more. The BJP controls everything today, and within that it is very difficult to get our voices heard, even in a flood. The people who came to distribute relief materials were also from the BJP, and they often gave us the materials at the end, mostly it was whatever had been left off after giving to everyone.

The process of hierarchical segregation makes certain communities reside continually at the receiving end of stigmatised social constructions which become a practical solution to everyday difficulties faced by the members of these groups. The 'working consensus' that Goffman (1956, p. 4) was talking about plays the role of a community peacekeeper, but in most cases, the peace benefits only the already powerful and dominant. Natural disasters such as floods displace people and communities, which have grave effects on their socio-economic lives. However, the effect of a natural disaster cannot only be analysed through objective analysis pertaining to incomes, material losses, and employment but also consider the

---

[9]This was shown empirically by the famous Sachar Committee Report on Muslims in India. The conclusion and summary of the report is available at: https://ncm.nic.in/home/pdf/recommendation/06-07.pdf (accessed 03.01.2024).

kind of emotions that people harbour within themselves for the artefacts and activities associated with them (Ge, 2019; Viju, 2019). Viju (2019), in his book on the ecological destruction of the western ghats (coast) of India, brings forward the effects that river flooding has on the various traditions and histories associated with a certain place. Under capitalism, the society has been facing graver threats of such disasters, where such disasters have become a process of marking individuals. This process of marking individuals straightens out individuals making them more amenable to social, economic, and political conventions determined by capitalism (Sethness & Clark, 2023). The growth of capitalism has meant that even environmental protection plans have begun to be evaluated on the grounds of the economic returns. Deshpande's (2018) work on floods in Mumbai, written from a corporatist perspective, is a testimony to the same. His work proves that for most corporates, economic growth holds more importance than the health, welfare, and safety of the people and the environment. Such an attitude complements capitalist urbanisation and planning processes that emphasise limited ownerships, uneven development, and greater economic returns (Harvey, 2005b).

The exertion of power and control by institutions is often found to be the primary force which keeps the social space in balance. Performances, put up by individuals, in order to avoid difficult situations which might arise out of inappropriate conversations or responses (Goffman, 1956, p. 4), are a result of the different codes and norms established by monumental institutions. A natural disaster breaks this stability in place. Gripped by fear and anxiety, many people resort to performing activities that they desist in normal times, as Chapters 2 and 3 have exhibited. The reactions of individuals to different situations are dictated by the codes which a society puts into place within the urban space. Under such conditions, Goffman's (1956) arguments about societies being theatres of performances comes alive. The performative spectacle, constituted by individuals, constructs the social reality as one knows it. Urban centres under capitalism are spaces where individuals lay their claim to belonging by taking the stage themselves, giving others the turn to 'act themselves' and secure their position within the space. The state of urban centres reflects Debord's (1968, pp. 4–5) theory about 'seemingly unconnected phenomena' contingent upon the social nature and construction of appearances, creating a spectacle capable of dominating human beings.

Capital's Routines and 'Natural' Disasters

The basic premise for tendencies such as petty-disaster capitalism emerging in the society comes from the lack of proper housing during a natural disaster. The residential spaces of people are the worst affected during floods, the destruction of which have both objective and subjective implications for the people (Ge, 2019). However, residential spaces are not the only life-sustaining places that a city builds for itself. They are a part of the larger corpus of built environmental features critical to the creation and sustenance of any contemporary urban space. While residential spaces provide a certain protective cocoon to individuals, built environmental features such as city halls, open spaces, and streets have been

118 'Natural' Disasters and Everyday Lives

known to be used by individuals and communities as communal institutions which are often quintessential to community integration process because they act as either 'buffer zones' or 'contact zones' (Pratt, 1992) or as both, contextualising the needs and desires of any diverse or multicultural community (Alexander, 1996; Kalka, 1991). The institutional nature of such buildings enables the peaceful and dialogic resolution of conflicts which is often found in spaces inhabited by numerous different kinds of ethnic, religious, or racial communities (Kalka, 1991). The *monumentality* associated with built environmental features helps in relegating the sharp boundaries and fault lines probe to generate conflicts, which are present in most urban centres (Suttles, 1973, pp. 25–26).

Built structures are not just mere concrete buildings, but they are, at the same time, 'pieces of art, political statements, cultural artefacts, [and] ... machines' (Ellard, 2015, p. 218). In other words, they act as symbols, which 'do more than merely stand for or represent something else [but they] also allow those who employ them to supply part of their meaning'. It functions in the classical ways of *monumentality* as Lefebvre (2003, p. 38) had defined it to be, that is being the generator of a paradoxical space where the monument 'extends far beyond itself, beyond its facade (assuming it has one), its internal space'. Such monuments become *supernaturalised* because they interfere with the interactional models of sociability creation of collective meanings and rules of different individuals and communities associated with them (Fine, 2021, pp. 7–8). These interferences occur in the form of directives, rules, and conventions, all of which are basic elements of any institution and constitute the dialectical intercourse between the institution as a body and its constituting members. They enable the construction of certain hegemonies through ideological manoeuvring similar to how Castells (1977) had argued, by making explicit the relations between the existent urban culture and broader socio-political and economic factors.

During a natural disaster such as a flood or an earthquake, thus, it is not only the built environment that gets affected but rather the actual effects are felt across collective memories, solidarity structures, and human cultures which get eroded by them. These aspects are critical to the formation of the lifeworld because these spaces provide the necessary avenues through which individuals engage with the space around themselves. In societies such as India, these spaces are more important because they provide, as Pratt (1992) and Kalka (1991) highlight, the avenues for individuals from different communities to come into contact with each other. The loss of a community space during critical junctures has grave effects for any city in India, which are often characterised by a wide diversity of ethnic communities, religions, and castes. For Barak Valley's population which roughly represents an equal division of Hindus and Muslims,[10] such spaces are more important because they provide a safe space for communities to discuss their issues outside the formal or legal process. Such built environmental features provide space to

---

[10]Such a population distribution has had an effect on the region's politics and social dynamics, particularly with the citizenship debates in recent times. See https://scroll. in/article/881600/in-assams-barak-valley-muslims-fear-the-new-citizenship-bill-will-disempower-them-politically (accessed 08.05.2024).

different communities and play an important role in inclusive community formation processes by providing a safe space to different communities, as well as help them in locating themselves psychologically into the broader community in place.

The effects of the built environment enable the formation of an everyday consciousness, which though dependent on the practicality of everyday life and remains almost always situationally unconscious (Lefebvre, 2002, p. 166), bearing a strong attachment or dependence on the monument. Such built environmental features in the urban space plays a large role in the reduction of conflicts and a creation of an urban culture. The sociologist Zygmunt Bauman described two dominant discursive notions of culture within the social sciences: 'one discourse generated the idea of culture as the activity of the free roaming spirit, the site of creativity, invention, self-critique and self-transcendence; another discourse posited culture as a tool of routinisation and continuity – a handmaiden of social order' (Bauman, 1973/1999, p. xvi). The general social order within any society functions along the latter, where social order is often ascribed great value. However, the social order that is prescribed to be in place by most is often highly exploitative for the marginalised populace of any social setting and further privileges the already privileged.

For the marginalised populace, disasters present an opportunity to assert their freedom within the society, often cast under the rubric of 'norm-breaking' and 'deviation' (Bauman, 1999, p. xvii) – manifested through routinisation and regularisation of lives. Routinisation of one's daily life such as fixing specific timeframes, establishing fixed norms about how a specific activity has to be performed, and the like plays a critical role in establishing the nature of the space from the user's perspective. High rates of routinisation ensure that people are often left with no time at their hands to reflect upon the daily occurrences unless somebody leaves some time to do that at the end of the day. For the people of the Barak Valley, the flood disrupted their routine daily lives, and as a resident puts it, 'gave [them] a huge amount of free time, with which nobody knew what to do'. These particular issues that people had faced during the floods had a class dimension as well. For the middle class and the elite, it meant a state of extended boredom, while for the marginalised, it meant a constant struggle for survival.

Social media becomes an important way to identify this contradiction in place. Contemporary society relies on social media for a variety of purposes, including one of community formation. However, as Deb Roy (2021) put it, these communities that people form are largely alienated in nature, which do not fulfil the classical role of the community but instead work towards entrenching people within the fold of capitalist ideological processes. Thinkers such as Putnam (2007) argued that ethnically diverse communities – such as India – are characterised by a certain lack of social capital within its residents in general because diversity, at times, creates problems in the formation and maintenance of social relationships in communities potentially leading to a kind of social withdrawal. This, in turn, affects the social cohesion within the space that they inhabit. Social capital also needs to be understood as an indicator of the amount of social cohesion these communities have in place (Fennema, 2004), in addition to the resources available to individuals of these communities through social ties (Small, 2009).

120 'Natural' Disasters and Everyday Lives

These communities, or social groups, are defined by the dual attributes of what they accept and assimilate as well as by what they reject (Lefebvre, 2002). The defining acts of communities in an adopted space reinforce the definition of place as being 'both a centre of meaning and the external context of our actions' (Entrikin, 1991). A key part of the above is the issue of the alienation that people face when they lose access to their ability to become productive or earn enough money to survive. Marx argued that capitalism produces the notion of productivity in such a way that the entire life cycle of an individual succumbs to the notion of productivity. Lefebvre (1991a), relating Marx's argument to the urban space that capitalism creates, put forward that productivity becomes one of the very foundations upon which individuals determine their level of integration with the society. Within such a situation, the creation of a routine and adhering to one's well-being itself becomes a challenge for many individuals, as many had stated during the floods:

> I have work. Yesterday, I had to put in a leave request on absolutely bogus claims. I feel guilty, but what to do? This place does not even have proper mechanisms in place through which I can work. I have to file in those requests. The major problem is there is no internet. They do not have any resources in place to provide internet during these tough times. I cannot even send some basic mails.

> The very fact that the entire household got so disturbed was that in my house, we have three people, out of the five, who were here for work for home purposes, which began during Covid. After that we did not go back thinking that we can save some money if we stay home. But then who knew that the flood would come.

The question of routines and schedules that cater to one's working life forms one of the basic aspects of contemporary urban life under capitalism.[11] Under capitalism, most of these practices are modified in ways in which they contribute to further alienation of the marginalised individuals. The routines that people build around themselves create avenues for alienation as these routines are usually constructed around the notion of catering to one's productivity as Lefebvre

---

[11]Simple daily practices such as walking are a multifaceted activity with effects felt right up to the individual's relationship with the minute elements of the urban space. It is the way in which individuals perceive the urban environment through the usage of their senses (Winkler, 2002, p. 8), actively or passively. Such experiences again are multi-sensory in nature and engage with almost all the senses within the body politic of the subject as Tuan notes, 'The five senses constantly reinforce each other to provide the intricately ordered and emotion-charged world in which we live' (Tuan, 1977, p. 11). Practices such as walking construct acts of dwelling integral to integrating an individual to the place and space (Ingold, 2011, p. 143). The natural disasters disrupt this framing of the community to most individuals.

(1991a) puts it. When people frame schedules, they do so purposively to situate themselves within the broader urban culture. The urban culture under capitalism encourages the creation of schedules because routines help individuals in mitigating the everyday possibilities of conflict and become a more effective part of the urban society. It is a critical part of the construction of an urban culture. The urban culture under contemporary capitalism is one that is fraught with contradictions. Culture, as it is understood in this context, is a shared set of beliefs about one's life in the city in which one lives through the social practices that one comes to embody within them and the social and personal relationships that they create (Fine, 2021, p. 10). Benjamin (2002) argued that capitalism produces routinisation and homogenisation, which becomes a tool to further the mechanisation of urban life. So, when people do specific activities at specific times, they actually create patterns for the urban space.

During the 2022 floods, such schedules or acts of belongingness were disrupted. The city's conceptualisation as an eternal form – which remains a capitalist manifestation of the segregation of the rural communitarian life from the urban life as Bookchin (2021) had argued – was put to test. The engagement of individuals with certain acts such as fishing and socialising with the community disrupted the capitalist scheme of routinisation whereby it becomes easier for monumental institutions to create certain notions of culture which overlook or tend to suppress the potential of the individuals to 'think outside the box' once the normalised ways of living have been established (Scott, 2009, p. 19). Natural disasters disrupt these routines that individuals form in their everyday lives, which, in turn, affect the ontological security that they construct for themselves through these routines.

The psychologist R.D. Laing coined the term, 'ontological security' to refer to the processes that allow individuals to experience their own subjective selves as 'real, alive, whole, and in a temporal sense, a continuous person' (Laing, 1969/1990, p. 39) emphasising self-identity and countering and alienating and altering environmental conditions that people find themselves in (Giddens, 1990, 1991). Ontological security is a measure of how individuals cope with change, both at the social and at the personal levels. Communities and local bodies play a crucial role in that. Co-operation and communities have been important parts of the human society, especially with regard to the marginalised communities. Floods disrupt the psychological constructions that people make with regard to a normal life (Lal, 2019). Jewett et al. (2021) have pointed out that during natural calamities, community resilience often comes to play a crucial role. But the critical problem is that community resilience in the contemporary society has become uncritically associated with a form of corporatist materialism that caters to the growth of an organic structure of capitalism.

Michael Lebowitz (2020) in his recent work has realised a dialectical approach towards analysing the relation between capital and community by focusing on the relationship between the whole and the parts. Communities under capitalism are classically understood as bodies where neighbourhoods manage the society and attempt to steer it in favour of the established norms in the society, promoting the domination that the commodity form has on the society. Commodities

122 'Natural' Disasters and Everyday Lives

under capitalism are not only valued for their use value but also because they produce a certain value. They are not just an entity focused on utility but are rather focused on the creation of a class conscious value form drawn from the dual nature of labour under capitalism (Dunayevskaya, 1958/2015). The commodity inherits the inherent contradictions within the nature of labour under capitalism and becomes 'in embryo [as an entity containing] all the contradictions of capitalism' (Dunayevskaya, 1958/2015, p. 85). The centrality of the commodity form which became evident even during the pandemic is such that any quantitative or qualitative change in their production, distribution, and consumption mechanisms greatly affects actual human lives. Capitalist production focused on the creation of profits does not only produce commodities but rather it produces the human being itself 'as a commodity, the human commodity, [a human being] in the role of commodity' in such a way that the human being only exists in the capitalist society in a completely dehumanised form' (Marx, 1844/1973, p. 284). The struggle that ensues within a capitalist society in terms of the challenges that a community faces is ultimately a struggle between the community and the commodity. Under capitalism, the commodities have a great advantage because they have the power to manipulate the idea of human needs (Deb Roy, 2021). Marx (1867/1976) describes commodities, as has already been stated, to be objects that are located outside of the human being and are characteristic of something that satisfies some form of human need. Marx argues:

> Objects of utility become commodities, only because they are products of the labour of private individuals or groups of individuals who carry on their work independently on each other. The sum total of the labour of these private individuals forms the aggregate labour of society. (Marx, 1867/1976, p. 165)

The domination of the commodity form creates the grounds for disaster capitalism and its associated process to grow within the society. Amid this, the marginalised sections of the populace in India often lack the voice and consciousness through which they can articulate the problems that they face. Natural disasters pave new avenues for these mechanisms and processes to take place in the society. The urban culture that capitalism developed in the society does not allow people to exhibit unalienated solidarity with others and instead outs the aegis upon the oppressed for maintaining civic order. This is highly problematic for countries like India where the kind of life that most of the marginalised and vulnerable populace live is one of the worst forms of human lives possible and is often dominated by a belief in supernatural or extra-human beings rather than one's material conditions. These beliefs do not allow the marginalised populace to generate a material consciousness that can lead to the acknowledgement of the scientific causes of floods and the exploitative framework therein. Exploitation is also the source of all forms of class struggle (Boswell & Dixon, 1993). The exploiter in a particular situation is accorded with some form of epistemic privilege which provides them with enough power to subdue the individual human subjectivity of those relatively powerless in comparison to them. This is the single most important aspect of the exploitative social reality that capitalism creates for people from the marginalised sections of the populace.

Disasters in India have become portals of opportunities for parliamentary politicians, for whom elections remain the primary goal, over and above the actual improvement of the conditions of life for the people. Such forms of reactionary politics fail to understand the importance of subjective policy framing that takes the entire diversity of the populace into cognisance. The individuality possessed by a person is a culmination of the interactions that people have with both other individuals and monuments. Institutions, when acting as monuments do, are to reduce the randomness – the indeterminacy – in everyday life by devising tactics which tend to homogenise the public space working towards developing a political space that cater to everybody. It reduces the multiple constellations of meanings which form the basic foundation of the social structure (Reckwitz, 2002, p. 255) and the social practices therein. Social practices define how human beings live. Habits define the processes in which human beings affect and get affected by the space around them. Andreas Reckwitz (2002, p. 249) defines 'practices' as being:

> [A] routinised type of behaviour which consists of several elements, interconnected to one other: forms of bodily activities, forms of mental activities, 'things' and their use of background knowledge in the form of understanding, know-how, states of emotion and motivational knowledge.

It is obvious that within an ethnically diverse space, these practices and their contextualisation within the urban space will be different for each individual and social group. Contemporary cities can be seen as manifestations of practices aimed at disciplining everyday lives (Allen et al., 2016). These acts of disciplining and control are performed at various levels, from the universal to the local. At the universal level, these are found creating global cities, while at the local level, they are found interfering with the processes of organic modes of cohesion and conflicts in urban spaces. A dialectical approach to the problem means seeing every process or part of the urban reality as being a part of the social totality (Kosik, 1976). This method encompasses within itself the conceptualisation of the dialectical struggle between the universal and the particular (Lefebvre, 2016b) – a struggle between the total and the local. Everyday social practices enable the development of social relationships of co-operation which can help in establishing not only community relations but also ideas about the politicisation of everyday life and routine activities. It enables the construction of an analytical framework which emphasises the continuous generation and perishing of social forms, relationships, and networks (Yates, 2014), especially in multicultural societies like India. Communities formed through these practices can be encouraged to make their own decisions and work through their differences, provided their dimensions and internal politics make them amenable to such processes. The underlying philosophy, in this context, needs to be that almost all the urban dwellers, in spite of all the differences they harbour, still have certain similarities within themselves especially in the context of their local economic and socio-political dynamics.

Chapter 5

Local Politics, Right to the City, and 'Natural' Disasters

Natural disasters often help people in coming together. Any place which has been hit by a major natural disaster has often exhibited great strides in collective reimagination of the society. Natural disasters – of the scale of the Kerala floods of 2018 or of the scale of the 2001 earthquake at Bhuj in Gujarat – have the innate capacity to generate senses of solidarity among people that go beyond the traditional contours of class, caste, and politics. For example, in a book on the Kerala floods, Anu Lal argues that the Kerala floods had brought together a great sense of solidarity and togetherness among the many different religious and political fractions of the state of Kerala. He writes:

> The amazing story of the creation of an egalitarian state, therefore, took the pilot's seat as the common cause – that guided everyone during the relief works. Therefore, the early days of the relief works were celebrated as the coming together of the people of Kerala from many walks of life in order to support and care for each other. This was another flood, a *pralayam* of love, a *pralayam* of caring, goods, money and workforce. (Lal, 2019, p. 12)

An egalitarian reimagination of any society requires interventions at both the central and the local levels. Local governance, in fact, has become an important part of the framework that the contemporary Indian state has been promoting with most political formations in India engaging with local elections far more vigorously than in the past. However, the engagement of most of these political formations has been within a neoliberal framework of governance (Deb Roy, 2024a). The inclusion of local administrative bodies into neoliberal frameworks has been one of the most critical aspects of the success of neoliberalism in India (Ahluwalia, 2002). The pattern of growth of the urban spaces follows the system of profit accumulation aided by neoliberalism that advocates in favour of the market gaining access to endless processes of accumulation of profits by employing various biopolitical and communicative mechanisms gaining extreme control

---

'Natural' Disasters and Everyday Lives:
Floods, Climate Justice and Marginalisation in India, 125–150
Copyright © 2024 by Suddhabrata Deb Roy
Published under exclusive licence by Emerald Publishing Limited
doi:10.1108/978-1-83797-853-320241006

126 'Natural' Disasters and Everyday Lives

over the people (Negri, 1982/1988). Amid the growth of importance that is being accorded to local politics by both the political formations and the corporates, tendencies such as municipal socialism[1] or local administrative egalitarianism have begun to make themselves more popular. In the context of India, municipalities are integrally connected to the communities because of the way in which the entire municipal governance infrastructure is structured in India. The community is an important aspect of urban life, and urban local bodies (ULBs), in turn, are an indispensable part of the community. The dominant community inevitably ends up controlling the municipal governance. The reports which came from the valley regarding the floods also reflect the disproportionate power relations in the area. In India:

> Municipalities are constituted by the State Government, which specifies the class to which a municipality shall belong in accordance with the provisions of the municipal Act. For this purpose, size of the urban population is the main criterion. However, in some States consideration is also given to other criteria, such as location of the urban area and the per capita income. (Aijaz, 2007, p. 10)

Municipalities play an important role in the process of decentralisation and in bringing in more effective forms of local democracy (Burns et al., 1994). The primary debates about local democracy and its importance in the political system were taken up by Joseph Chamberlain, one of the major proponents of the civic gospel in Europe, who believed that the focus on local issues such as the democratic provisioning of gas, water, and housing, and their correct resolution is the fundamental element of all kinds of egalitarian politics (Gehrke, 2016).

The focus on the local administrative unit and its functioning was one of the basic aspects of India's pandemic response and continues to remain an important plank of all kinds of disaster-mitigation policies. However, local administration and politics has been particularly affected by the growth of non-governmental organisations (NGOs) in the society which have often tended to take up the role that was previously entrusted with the local administration. The current chapter takes cue from this position and discusses the implications of local politics in three sections. The first section talks about the localised nature of the NGOs during a natural disaster and the class dynamic associated with them. The second section speaks in terms of the communitarianism that developed during the floods, and the third section relates the previous two sections in arguing about

---

[1] Municipal socialism can be defined as a system through which the socialist politicians and activists tend to mould the municipality – an important part of the community as has been spoken about in the previous chapters – in accordance with their overall socialist views (Jowett, 1907). The idea of a localised welfare state was used as a foundation for the conceptual formation of the theory of municipal socialism by the liberal political figures of Manchester during the mid-19th century (Bönker et al., 2016).

Local Politics, Right to the City, and 'Natural' Disasters **127**

the centrality of the right to the urban space that the marginalised people need in order to make themselves a part of the urban space and the ways in which a natural disaster affects the same.

Localism and NGOs

The significance of local politics during a natural disaster increases prolifically because some of the issues that emerge during a major natural disaster are highly localised in nature and demand different kinds of solutions than the ones that a centralised bureaucracy produce and conceive of. However, the extent of success that a localised mode of resistance can generate also depends upon the ways in which local politics relates itself to the broader national and global political dynamics (Singh, 2009). During the 2018 floods in Kerala, the 1,200 vibrant local self-management and self-government structures put in place played a crucial role in mitigating the effects of the floods (Rajamony & Mana, 2022). These 1,200 institutions included 941 gram panchayats, 152 block panchayats, 4 district panchayats, 77 taluks, 87 municipalities, and 6 municipal corporations. Assam, on the contrary, has only around 105 ULBs in total, along with 2,197 village panchayats,[2] which is abysmally low considering that Kerala's population is 3.46 crores and Assam's is 3.66 crores (around 34.6–36.6 million).[3]

The ULBs have a tremendous ability to mitigate the latter kind of issues. The local bodies, operating through the communities, can become the saviours of the marginalised if they are democratised in such a way that they do not operate through the monumental effects of the state, the market, or the community. The growth of neoliberalism, however, has made it difficult for the ULBs to perform these requisite duties. The ULBs, out of compulsion, instead of advocating for greater control and resources had to resort to further exploiting the marginalised workers. One of the key reasons for the same being that most of the ULBs have been subsumed under the growth of neoliberal assertion within the domain of local politics, policy-making, and infrastructural issues. The range of problems that urban governance faces in India is multifaceted, which span from financial incapability to administrative inefficiency. The per capita expenditure of municipalities in India, as of data from 2012 to 2013 financial year, is estimated at Rs 3,116. This also includes the per capita revenue expenditure and capital expenditure of INR 1,986 (20 GBP app.) and INR 1,130 (11 GBP app.) respectively (Mohanty, 2016). According to Bhavsar et al. (2020), the uneven and often disproportionate allocation of resources across different municipalities in the country has resulted in municipalities being unequally equipped to engage in carrying out their supposed functions, (Ahluwalia, 2019).

---

[2]Data collected from https://localbodydata.com/urban-local-bodies-list-in-assam-state-18#google_vignette and https://localbodydata.com/zila-parishads-list-in-assam-state-18#google_vignette (accessed 02.03.2024).
[3]See https://censusindia.gov.in/census.website/data/population-finder (accessed 06.04.2024).

128 'Natural' Disasters and Everyday Lives

The tasks that had been once ascribed to the ULBs are increasingly being handed over to the NGOs. The replacement of local politics' importance by the NGOs in the mainstream Global South means a replacement of the economic functions that local bodies perform. In the South of the Global South (SGS), however, it means something much more than that; the dominance of NGOs and their associated liberal media means that the biopolitical control that characterises the society gets further entrenched within the overall social psyche. Hardt and Negri (2004) had highlighted that NGOs (and their liberal counterparts) play a crucial role in the generation of biopolitical control of the society because they act on behalf of corporates and capitalism in general who are the basic source of their existence. The Marxist scholar from India, Randhir Singh, writes:

> [W]hile the work of best of [the NGOs] is indeed deserving of admiration, it is a mistake to look for answers to the environmental crisis, as not a few have done, in the local initiatives of voluntary activism of the NGOs. Taking into account such serious issues as sources of funding, vulnerability to manipulation by vested interests, coverage and development of initiatives and innovations, record of overall achievement so far, and the limitations inherent in local initiatives where basic issues of economy or politics are involved, there is little to share optimism about the effectiveness of NGOs in reversing the current trends and protecting the environment. (Singh, 2009, p. 41)

While Singh's argumentative position against NGOs is an important one, the problems are more complicated in the SGS. In the SGS, the issues are much more socialised in nature because the community structures still continue to exert an effect – that is much greater than their counterparts in other areas – on the kind of social reality that the marginalised face. The choices that the marginalised make (or have to make) are determined by a combination of their social reality, their social perceptions, and access to economic resources, which together shape their class aspirations. Aspirations and desires enable the neoliberal social structure to produce de-politicisation and atomised individuals by focusing on personal achievement, meritocracy, and individual responsibility. These factors during a disaster such as a flood become barriers in the creation of a solidarity-based social structure in place that can bring forward a more egalitarian approach towards disaster management at the community level.

The middle classes get easily attracted to Civil Society Organisations (CSOs), because CSOs address issues that they can easily identify with in terms of their lifestyle and aspirations (Sen & Dhawan, 2015). The questions of lifestyle become important for the middle classes because they do not confront questions of survival on an everyday basis like their counterparts from the highly marginalised social groups. To strengthen the capacity of human societies to engage with disasters, one needs to focus on, as Brownhill and Turner (2019) argue, the creation of a subsistence-based post-fossil-oriented social movement with highly marginalised sections – more so, the women among them – asserting themselves both socio-economically and politically. Small communities within peri-urban regions, as Wall (1999) note, are much more effective in such collective organisational forms of resistance than larger urban spaces. If one follows the work of Henri Lefebvre, one can get a sense of the implications behind such a shift. In metropolitan spaces, the human being's role in the space is mostly determined by the extent of productivity that one can generate (Lefebvre, 1991a), coupled with the complete subsumption of individualities to overtly abstract formulations.

The marginalisation of local politics often means a marginalisation of the concrete contradictions that people face in their everyday lives. It has often come to mean that abstract and more universal contradictions are preferred over and above the everyday issues that people face which contributes to a gradual dismantling of everyday concerns benefitting capitalism and authoritarianism because everyday concrete struggles are the basis of the people's movements against exploitative tendencies (Soper, 2020). The lack of a local focus often results in a growing alienation of the people from the actual problems that they face which has implications not only for the social cohesion in place but also for universal aspects such as democracy (Sullivan, 2020). The complete focus on the claims that the rupture of the dike was the main cause behind the floods contributes to this process in Silchar. The focus on the dike and narratives like 'flood jihad' made the entire discourse restricted to a spectacle. Within the theoretical contours of disaster capitalism, such a focus creates the adequate grounds for neoliberally oriented political consensus to bring in further othering in the society. However, the solution to such a biopolitical mode of exploitation cannot be based on the creation of a corporatist model of resistance and advocacy that most NGOs bring forward. It is important to remember as Arundhati Roy argues:

> NGOs give the impression that they are filling the vacuum created by a retreating state. And they are, but in a materially inconsequential way. Their real contribution is that they defuse political anger and dole out as aid or benevolence what people ought to have by right. They alter the public psyche. They turn people into dependent victims and blunt the edges of political resistance. NGOs form a sort of buffer between the *Sarkar*[4] and public. Between Empire and its subjects. They have become the arbitrators, the interpreters, the facilitators. (Roy, 2014, para 4)

A policy focused on NGOs often tends to ignore the possibility that local politics can make a significant impact upon the mitigation of the effects that a disaster has. The relevance of local politics is significant in the context of disasters such as floods, which displace communities and make it necessary for people to make changes to their regular schedules and routines. Speaking on the topic, two residents said:

---

[4]The Hindi term for 'Government'.

130 'Natural' Disasters and Everyday Lives

> During the floods, we cannot go about normal lives. Normally I used to get up very late in the morning. But during the floods, because there was no electricity and no internet – that means no videos or scrolling at night – I used to wake up very early. That was such a mess. After a certain point of time, it was so frustrating. The entire day, there was nothing better to do, just look at the water and see the boats.

> The floods disrupted work. For me, I have to wake up early, very early at around 4 in the morning, so that I can go to search for work. With the floods now, I have nowhere to go. I do not feel like a human anymore. I need money to live like a human being in this word, and for that I need to get into my old routine because that routine had served me well for so many years. The camp is ok, but it does not give employment which is important part of what you refer to as being home.

Despite the statements made by the two seemingly being from completely different class and caste backgrounds, they still reverberate the lack of belongingness that they feel towards their 'new homes' – the first one referring to a home without electricity and internet, and the second referring to a relief camp – as the first and second respondents above further describe:

> After the floods, we cannot even recognise our own homes. There has been so much water in it that even the tiles have broken down. Now we do not have that much money to redo the tiles and basically have to wait for another two to three years before we can land our hands on that.

> Our home is washed away completely. There is no place to go back to. Along with that, all the utensils and some clothes also got washed away including my daughter's school uniform. It will take months to get back the old things. God knows when we will be normal again.

Such issues rupture the protective cocoon of familiarity that individuals build around themselves. The generation of the protective cocoon is contingent upon the notions which an individual harbours within oneself regarding the concept of home and its associated identity construction processes (Dupuis & Thorne, 1998). The regeneration of such cocoons requires a local focus, one that is incapable of being generated by strategies such as the usage of NGOs and community volunteers only but rather requires a widespread reorganisation of the society itself. The changes in one's regular lives often have a distinct effect on how they engage with the society in general, especially when the perception that they have towards certain activities undergoes a massive reordering. Activities such as fishing and gossiping, which otherwise might have been categorised as being the remnants

Local Politics, Right to the City, and 'Natural' Disasters   **131**

of a bygone era, become the only avenues that most of the people have towards spending their time. In the absence of electricity, mobile connectivity, and internet, chatting with one's neighbours had become the most important thing that one could engage in to get news of the world. As Amina, a flood victim,[5] put it:

> It was hot, with high humidity, but at the same time, it was also quite sublime. No electricity, no internet, it was tough but sometimes, it was also a reminder that these things are just things, and that we can live without them as well. It is childish, I know, but I took to making paper boats, and throwing them into the water. I was also doing a lot of things in the water. My mother needed help in getting the water from downstairs, so all of these activities I did. It was nice to reconnect with them once in a while, and do activities together, like we used to. The major issue was getting the water from the pump. We have a two storied building, so we have water, but the tank went under water. It was a mess getting the water. Our tank was under water, but the tap was still above the water level, so we had to get a long pipe with which we filled up the buckets with me standing in three feet's of water. It was an arduous task, but I had to do that as nobody else could have done that, not my aged parents at least.

Under contemporary capitalism, activities such as pulling buckets of water by hand are considered to be unproductive and sub-human because technology has mostly taken these activities over performing them at a shorter time and with a higher productivity. Such activities – often associated with unproductivity and a waste of one's time – become the lifeworlds of the affected people. For example, during natural disasters such as a flood, fishing fulfils two important communitarian functions, the first one being the provisioning of food (in the form of fish) and the second being its ability to make people connect with each other at a more personal level. Such activities help people in rediscovering a sense of routine in their lives and a sense of place. Most of these factors lead to a rise of community bonding as well, which goes beyond the mainstream profit calculus that dominates the contemporary society. Kuttappan (2019) in his book on Kerala floods narrated about the service – often unpaid and unacknowledged – that the fishermen of Kerala gave to the community, which played a large role in the mitigation of the effects of the floods. Such steps can offset and counter the institutional lethargy that characterise the response to floods in many of the disaster-prone regions in India which often increase social conflicts and erode social cohesion during disastrous events.

The right to the city is about negating the process in which individuals lose the connection to the urban space which produces a notion of being lost without a community, which results in individuals '[recognising] their areas less clearly, evaluate them less highly, and become less attached to them' (Hunter, 1974,

---

[5]Name changed on request.

132   'Natural' Disasters and Everyday Lives

p. 173). Being engaged with the community, albeit with certain restrictions, makes people appreciate the communitarian way of living, which is frequently threatened under the neoliberal corpus of social restructuring based on heightened productivity and toxic work cultures that create an alienated conception of the protective cocoon within individuals. The routine which an individual constructs, and the social practices enmeshed within it, is important for the continued sustenance of the individual's 'protective cocoon' (Giddens, 1991, p. 56) and is also as much a physical activity of boundary construction as much as it is a mental activity of interpreting and creating an interactive framework with social reality (Reckwitz, 2002).

Under contemporary capitalism, however, this social reality is constituted by a range of consumer artefacts that produce the human individuality and subjectivity (Singh, 2009). While the initial years of independent India focused on the creation of a manufacturing sector, the 1980s were a period dominated by partial neoliberalism where there was an encouragement to produce more consumer goods (Rani & Unni, 2004). The growth of consumer goods has resulted in a situation where many such commodities have become a part of the everyday life of the people. Aspects of contemporary lives such as mobile phones, electricity, and gadgets have now become essential parts of urban life and a crucial factor that determines the perception that one possesses regarding the space that they occupy. For example, during the floods, there were a few commodities whose sale had increased manifold. Most of the sales were driven by the lack of electricity and mobile connectivity. These commodities included generators and motors. In the contemporary times, factors such as electricity and internet have become such an integral part of the daily lives of people. Two flood victims, Akash and Rajesh[6] state:

> During the floods, with the electricity off, there are only a few things you can do. So, I took to fishing. I used to do that a few years ago when we still had our pond. But then with the building coming up, we lost that. So, I could do that again with water everywhere around me.

> The basic issue was mobile connectivity. That was not there, and so was no internet. What do you expect? I had to talk to my neighbours, people with whom I have not spoken to in a long time. These connections, I guess are what makes us human. It helped me regain a lot of my old friends as well, although I have lost contact with them again.

Different individuals have different kinds of requirements. While for Akash above, the basic issue was electricity because his house was just too warm for that kind of a weather, for Rajesh, the issue was mobile connectivity because he was struggling to come to terms with a life devoid of social media. The kind of relations that most individuals share with the contemporary society is dictated by technology, which has caused a rapid rise of alienated communitarianism (Deb Roy, 2021). The social relations that individuals share within a community are

---

[6]Names changed on request of the interviewees.

Local Politics, Right to the City, and 'Natural' Disasters **133**

dictated by the motives and expectations that a community possesses. Domestic workers become a part of an urban community not only because of the objective aspects of the work that they do but also because their work directly intersects with the subjective aspects of a person's class consciousness. When such activities get disrupted, there is a rupture of the basis of belongingness, which is usually constructed out of 'glass palaces' which ensure the psychological security of the individuals, which get easily destroyed during floods when people find themselves stripped off the benefits of modern everyday life.

The lack of social media had played a critical role in the worsening of the conditions faced by the victims. It has been acknowledged that the kind of community that one forms on social media is often an alienated one that is stripped off the radical possibilities that a communitarian structure possesses (Deb Roy, 2021). The floods disrupted the workings of such alienated community formations and provided a window towards more participatory communitarian practices in the form of increased human interactions and a respite from the heightened productivity-based functioning of the social factory. However, the lack of social media has had other implications as well. Ge (2019) and Kuttappan (2019) have narrated how social media helped the authorities and other community activists in rescuing victims during the Chennai and Kerala floods. During the Silchar floods of 2022, however, the absence of social media was felt acutely with many residents – in the absence of internet or mobile connectivity – unable to call anybody for help except their immediate neighbours and community members.

Many people had to shift to radio because of it being independent of electricity requirements – which again forced many to buy batteries priced three times higher than their usual prices – to know about the water levels. The critical role that social media plays in mitigating the chaos produced by rising water levels was felt during 2024 when social media played a critical role in letting the people know about the flood levels through social media websites and platforms such as Facebook and WhatsApp. This was particularly helpful for those who could not flock to Annapurna Ghat – as can be seen in Fig. 22 – to check the constantly changing water levels at the display board at the local flood cell office (shown in Fig. 23). As one resident puts it:

> They say that we should not go near the river, but we depend on the river for survival, for water, and for fishes which we eat and sell. We cannot afford but not go. If they had been serious, they would have put in security measures, or gave us some procedures to follow, but nothing of that sort was there. They just said, do not go which we cannot afford to do.

Some of the issues raised during the floods can be more effectively resolved with a more radical local administrative set-up. However, the effective and democratic management of any city involves institutionalisation and public accountability of the administrators, politicians, and others in positions of power. This is extremely difficult to be achieved within a highly bureaucratic and authoritarian system in place. Local elections in this regard proved to be critical because they provide an opportunity to advocate for a greater control of municipal resources

Fig. 22. People Flocking to Annapurna Ghat to Check Water Levels.
Photo credit: Author.

Fig. 23. The Flood Cell Office and the River Water Display Board at Silchar.
Photo credit: Author.

to the community. Municipal resources can be directly utilised by the community because under ideal circumstances, the community can effectively control the operations and finances of the municipality. Under neoliberalism, however, the municipality and its associated resources begun to get alienated from the community because the matrix of domination that operates within the society does not consider the community to be an important part of human existence. It seeks to replace the communities by NGOs, which enable it to further its own biopolitical domination.

Communitarianism, Municipalism, and 'Natural' Disasters

The first formal local elections *as one knows it today* were held in 1889 for county councils in Britain, where the 'principle of representative local democracy' was highlighted by the contesting and voting population (Barron et al., 1991, p. 1). Since then, local governance has taken on a diverse path globally, with a highly fluctuating importance being given to it in various political and social cultures in a variety of different social socio-political systems. However, with globally increasing urbanisation, the ULBs have again found themselves to be at the centre of attention. Urbanisation under capitalism has increased the propensity that citizens have towards experiencing disasters, not only because of the changes in the environmental balance but also because of the changes in the social needs of individuals.

Aspects such as social media, internet, and electricity, which became a major issue during the floods, are parts of the socially constructed need-based framework that caters to the populace. They become a part of the consumerist framework that capitalism creates, and the psychological effect of their absence creates a further demand of these entities in the lives of the people. After the floods, this was more accurately visible with a great spike in the usage of social media and electricity consumption with more people engaging in debates and discussions over the implications of the floods, reminiscing about the horrors of life without mobile connectivity, internet, and electricity. Needs such as this today, however, are socially structured and created, and these needs are manipulated by the capitalism which work towards the creation of newer needs (Hardt & Negri, 2004). The private space, which occupies a pivotal position in the life of any individual, regulates and gets regulated by the individual's social practices within the social space. The relationship between an individual and the community depends upon the mediations taking place between the individual's needs and desires and the social relationships already existent within the social space of which the space of the community is also a part of. Certain needs, which are integral parts of everyday lives, such as leisure, become commodified and alienated from the everyday life within capitalism (Lefebvre, 1991b). Needs, however, are different from desires because under capitalism, 'desires no longer correspond to genuine needs; they are artificial' (Lefebvre, 2002, p. 11). The provision and means to fulfil these needs, however, have been appropriated by neoliberal reforms.

Electricity, internet, and mobile connectivity systems today increasingly depend upon a growing contractual workforce that is employed on a piece-rate basis. With increasing contractualisation, the responsibility of the state towards the workers has decreased. At the same time, it has also meant that the state has

136 'Natural' Disasters and Everyday Lives

lesser control over the provisioning of these services. Contractual workers often do not possess the sense of responsibility towards the society like their permanent counterparts. Most of them do not do the one single job but also do a myriad of other jobs to sustain themselves. During the floods, many of these contractual workers – mostly from poor socio-economic backgrounds – took up other jobs to sustain themselves, as one of them stated:

> The lines were down, and I am paid piece rate only. So, why will I risk my life for this job. I gave a shop near my house on the first floor of a building. Got some 2-3 customers every hour, it was good business. If I had been a permanent worker, I would have probably worked, but I am not risking my life for INR 300-400 (3-4 GBP app.), when I can do business for INR 500-600 (5-6 GBP app.) every 2-3 hours.

The low availability of workers has also been impacted by ideas coming from neoliberal scholars who have argued that contractualisation can indeed be beneficial both to the employee and the employer because it not only cheapens the price of the service but also provides more opportunity of personal growth for the worker (Mohan, 2015). In reality, privatisation and contractualisation of local municipal services is actually related to the growing fiscal conservatism that has gripped many local bodies across the world (Ascher, 1987). Outsourcing of essential services and commodity provisioning has often had a disproportionate effect on their accessibility for the marginalised people. For example, during the floods, the administration used many scheme workers for performing relief works (District Disaster Management Authority Cachar (DDMAC), 2022a), most of whom had no experience in such activities. This points towards a general sense of negligence in the society, as far as the essential service provisioning is concerned.

With the outsourcing of essential services such as sanitation and public distribution of essential goods, the administrative duties that an officer or a representative must perform have become extremely complicated under neoliberal capitalism. Many of them, as the following statement reveals, were completely at their wits end when they had to respond to the floods because they neither had the expertise nor the autonomy to do their duties in a timely manner. This is in stark contrast to the traditional role of the state in India. Since the ancient times, India has had a system of administrative duties being performed by officers, who were often bureaucrats from the middle classes. Because of their class nature, most of them could never really come to terms with the actual problems of the marginalised. The constant inclusion of middle-class individuals in positions of power – even if they come from vulnerable and marginalised backgrounds – has done little to change the nature of local administration in India, regardless of the kind of ULBs.[7]

---

[7]In India, there are different kinds of ULBs that function at different levels, such as municipal corporations, municipal councils, and nagar (towns/cities) panchayats, all catering to different quantitative samples.

Local Politics, Right to the City, and 'Natural' Disasters   *137*

The one commonality between all of these different kinds of ULBs was that most of them have been starved of funds and financial autonomy. The neoliberal assault on their status as decision-making bodies has made most of these ULBs powerless in the face of the commercial assertion by the market. The ULBs in India – the ones entrusted with local administration – had to perform some of the most important duties during the pandemic. However, in spite of that, their financial autonomy remains highly constrained. This constrained nature of the ULBs' financial fabric often results in them not being able to perform the duties that they are supposed to perform to the best of their capabilities. In the context of India, ULBs are not only important because they are the lowest, often called the third, tier of the government but also because they are a part of the community they represent, much like the panchayats in rural India. Municipalities are one of the most important parts of the system of urban governance in the country. They are essential elements of how people like Thomas Attwood and Joseph Chamberlain conceptualised 'town councils as real and legal political unions ... which would give the people a better means of making their power felt' (Fraser, 1987, p. 34). The possibilities of local governance have even caused neoliberal thinkers, such as Gurcharan Das (2002/2012), to praise the promotion of local governance by states actively embracing neoliberalism.

Neoliberal intervention into the everyday lives creates the conditions of the establishment of the biopolitical control that capitalism desires to implement in the society. According to Lefebvre (2008, pp. 84–86), there are three factors which intervene in the construction of everyday lives: factors of homogenisation, factors of fragmentation, and factors of hierarchisation. Within this framework, the analysis of local communities as a complex web of 'interactional networks, shopping practices, and local usage patterns' (Suttles, 1973, p. 12) becomes an extremely useful tool to the study of social and communal cohesion because it reveals the holistic analysis, encompassing the positive and negative aspects of everyday lives within a community. As Lefebvre says:

> The city creates a situation, the urban situation, where different things occur one after the another and do not exist separately but according to their differences. The urban, which is indifferent to each difference it contains, often seems to be as indifferent as nature, but with a cruelty of its own. (Lefebvre, 2003, pp. 117–118)

Tendencies such as municipal socialism or democratic and libertarian municipalism take these issues into account. Such terms have been used to refer to various forms of local-level socialist planning initiatives, including, along with 'municipal socialism', aspects of 'city socialism' (Rustin, 1986), progressive municipal enterprises (Cohn, 1910; Palmer, 1986), and so on. Municipal socialism emerged as a strand of the broad array of radical alternatives to civic capitalism, emphasising the advantages of municipal ownership and urban councils' borrowing powers (Gehrke, 2016, p. 26). Libertarian municipalism, on the other hand, as advocated by Bookchin (2015), argues in favour of a more radical form of municipal governance that speaks of more direct forms of democratic management, one where

138 'Natural' Disasters and Everyday Lives

the political formations remain under the control of grassroots organisations. Put broadly, these forms are a multifaceted praxis-based socio-political theory constituting 'a set of roughly analogous historical movements that fought the private delivery of essential services and used local governments to advance "socialist" agendas', which encouraged increased sector employment, collective universal provisioning of utilities (Leopold & McDonald, 2012, pp. 1837–1838, 1843–1844), economic renewal, employment, socially necessary production (Rustin, 1986, pp. 75, 79), and so on.

However, most of these tendencies often work through a normalisation bias, where the urban middle class forms the basis of their theoretical and practical political praxis. To generate an effective belonging to an urban space, the marginalised need to possess the right to produce forms of cultural commons which is, as Hardt and Negri (2009, p. 350) argue, 'dynamic, involving both the product of labour and the means of future production'. Such a conceptualisation of the common includes the languages that society creates, the social practices therein, and the varying models of social and personal relationships which collectively and in a non-commodified sense 'establishes a [social practice and] social relation ... whose uses are either exclusive to a social group or partially or fully open' (Harvey, 2012, p. 73). Harvey (2011, pp. 103–104) notes that it is the social practices performed by individuals and various social groups within and through the diversity that an urban city space has to offer, which constitute the human quality characteristics of a city space. These practices are the essential foundational blocks of a common culture, within which all the inhabitants of the city and the city itself dwells in. Individuals in such spaces enter into specific socio-political relations which they have to maintain, even if they have not chosen them or depended on them (Lefebvre, 2016a, p. 32).

The private space of individuals is also characteristic of the way in which the individuals negotiate their relationship with the broader social dynamics in place (Marx, 1844/1973). That being said, it is important to note that the social relations that people form under capitalism are almost always detrimental to a certain section of the populace on the basis of class, caste, race, or gender. For example, during the 2022 floods, many women complained that their responsibilities within the household increased manifold regardless of their class position, as two women – the first from a relief camp and the second staying at the second floor of her house – stated:

> In the relief camp, the household work increased manifold. At home, I could plan it, but here it is very difficult. Most of us women often do household chores when the men go to work, but here in relief camps, it is not possible since most men do not have work anymore, and they just loiter around. They often drink and are wasting money, quarrels and fights are also increasing. Household work has increased also because the children are not going to play, and I have to take care of them all the time.

> It might look that being at home, life is easy in this deluge, but it is not so. My husband does not do any household work. During normal times, it was never a problem, but with water and all, it is becoming difficult. Going knee deep in water every day to get stuff from the cupboard downstairs to eat has become a difficult job. Water is also an issue; we have to drag that up and then boil it and then drink. All this is to be done by me. That is extra work.

Such conditions of life are not alien to India's hinterlands. In relief camps – like the society in general – women are usually exploited more than men, and additional care for women becomes necessary because of varying biological and social attributes possessed by women (Hirani, 2022). These issues affect the quality of everyday lives that people face within relief camps, especially if they belong to vulnerable sections of the society. The everyday lives of individuals are the most critical sites of humanisation (Lefebvre, 1991b, p. 163) that takes place within a complex circuit of needs and desires, which supersedes the way in which Tönnies theorised the creation of a community based on a certain moral basis reflecting family ties, land ownership, social norms, and mores (Giuffre, 2013, p. 20). Individuals within urban spaces frame notions of being secured (or insecure) through their everyday practices and interactions which affect and are affected by the urban space around them, and relief camps are no different as well.

Relief camps frequently become sites where existing psycho-social issues manifest themselves during a disaster fuelled by the coping mechanisms that individuals and communities exhibit during such conditions (Amarakoon et al., 2024). For the women mentioned above, the community that they became a part of during floods reflected a highly deplorable state of existence, whereby their ability to mark their boundaries which act as markers of their social identities and social ordering (Rothbart & John, 1985; Tajfel, 1974) was significantly reduced. It is in this context that many of them went on to become manifestations of petty-disaster capitalism because it allowed them to escape the same and move towards newer forms of community engagement. The most critical aspect of capitalism in this regard remains its success in ensuring that the terms that the women found most easy to engage themselves with the society – to counter their social and economic marginalisation – came through their practice of petty-disaster capitalism. Local authorities remained inept at navigating through these changes because most of them, as Shaw (2005) argues, in India have been stripped off the autonomy that they possessed, and autonomy, as Stromquist (2023) argues, remains one of the most important parts of the radical vision that local governance promotes in the society.

Disasters and the Right to the City

Cities are not created out of thin air but are rather manifestations of the social fabric including the 'lifestyles, technologies and aesthetic values' which exists among the people who design and live in the city (Harvey, 2008, p. 23). But under

140 'Natural' Disasters and Everyday Lives

the influence of monuments and the social culture which builds up around monuments, individuals often do not create their own identities but rather conform to already existing cultural norms and prefabricated consumer culture, especially under conditions of contemporary forms of modernity (Blackshaw, 2005, p. 84).[8] The security of ontology can only be realised realistically if the marginalised are granted the right to the city which effectively means having the ability and potential to 'change it after [one's] heart's desire' and not merely a right to public services and access to resources (Harvey, 2003, p. 939, 2008, p. 23).

The state of the publicly managed resources occupies an important position in the analysis of the impact of neoliberalism in the society. While it is true that the existence of the public sector is crucial to any developing economy, the mere existence of the public sector does not mean that the services will be distributed in an egalitarian fashion. The public sector in any society is merely a manifestation of the ideology and nature of governance that one finds in a particular social structure. However, it is important to consider that the state – as Chapter 1 shows – cannot be taken to be a revolutionary force in itself. The revolutionary force of transformation occurs through the events in one's everyday life, rather than the monumental institutions that one is a part of. Social cohesion and social conflicts occur within the everyday lives at the local level, and it is this level of the social whole which becomes critically important in the envision of a social revolution aimed at complete de-alienation of the everyday reality of capitalism (Lefebvre, 2008, pp. 15–16; Vaneigem, 2012, p. 11). Everyday lives are immensely powerful locations of analysis if one is to envisage social change in a radical manner (Goonewardena, 2008, pp. 117–121). But everyday lives are also highly indeterminate in form and content. This is because 'Many men, and even people in general, *do not know their own lives very well, or know them inadequately*. This is one of the themes of the critique of everyday life' (Lefebvre, 1991b, p. 94). Capitalism, as Lefebvre (1991b, p. 157) argued, produces individuals who do not have any right over the various objects which constitute the urban space and the society in general.

The concept of the *Right to City* is the process through which a variety of mechanisms is put into place by which the marginalised populace in any urban centre can assert their right over the urban space through a variety of parliamentary and extra-parliamentary mechanisms, which include social movements and changes at a policy level by political participation. In simpler terms, the Right to City concept argues that cities and urban centres are constructed keeping in mind the elites who inhabit and go on to control the city space. For marginalised people to be an effective part of the urban space, it is necessary that they possess the right to alter these spaces. Lefebvre (1996) analysed these acts of remaking as

---

[8]Under 'Liquid Modernity' a form of modernity theorised by Bauman (1999), the identities of individuals, as Blackshaw (2005) argues, the identity of any individual is bound to be dependent upon existing forms of culture because liquid moderns are easily absorbed by existing norms with little potential for structured deviance.

Local Politics, Right to the City, and 'Natural' Disasters   *141*

a part of the struggle towards exercising the '*Right to the City*'. In other words, as Harvey says:

> The cry [for the *Right to the City* by Lefebvre] was a response to the existential pain of a withering crisis of everyday life in the city. The demand was really a command to look that crisis clearly in the eye and to create an alternative urban life that is less alienated, more meaningful and playful but, as always with Lefebvre, conflictual and dialectical, open to becoming, to encounters (both fearful and pleasurable), and to the perpetual pursuit of knowable novelty. (Harvey, 2012, p. x)

A true sense of ontological and material security can be evoked through a manifestation of the right to the city that marginalised people inhabit. In the words of Andy Merrifield, the right to the city is a concrete right which 'means the right to live out the city as one's own, to live for the city, to be happy there' (Merrifield, 2017, 'Fifty Years on: The Right to the City'). The right refers to reinvigorating the notion of belongingness, which can only be truly realised if the social totality retrenches itself from the dehumanising impacts of neoliberal capitalism which homogenises and routinises life. In other words, the autonomous model of community formation, the search for shared solutions to collective problems (Bauman, 2001b; Beck & Beck-Gernsheim, 2001), and the realisation of collective outcomes of urban social processes (Suttles, 1973) will have to be reinstated. The obstacle to these realisations is the rendering invisible of the individual within the society as a subjective agent – whose autonomous agency needs to be re-established by processes of de-alienation. The urban way of life, as Monti (2000, p. 21) says, is 'not something physical like a tall building or a bridge. Nor is it as familiar to us like our family, neighbourhood, or job. It is more like a set of public habits or customs'. It is the responsibility of the people residing in urban spaces and cities to interact with the established civic culture composed of urban rules, rites, and norms – alter them if required and ensure their intergenerational transmission (Monti, 2000).

Speaking from the context of Assam, Das and Mitra (2003) have highlighted that the structural measures need to be accompanied by non-structural measures as well, such as disaster relief, public health measures, flood insurance, floodplain management, and increasing disaster preparedness. However, in doing so – like most of their counterparts – Das and Mitra take a highly liberal standpoint which does not focus on the global politics of climate change but instead reduces the flood problem of Assam to a local one. Scholars such as Singh (2009) and Narain (2017) have argued that climate justice requires a universal effort from the developed and underdeveloped regions, with a specific focus on the development of sustainability in underdeveloped regions of the world. It is important that one comes out of the arguments advanced in favour of a flattened world where climate change is seen to be a universal matter affecting everybody equally that can be countered by a highly globalised industrial model (Foster, 2009). Notwithstanding the failure of capitalist and market-oriented governments across the

142 'Natural' Disasters and Everyday Lives

globe, there do remain a tremendous number of individuals and organisations who continue to believe in the radical and optimistic capacity of capitalist economic, political, and social processes. They often fail to realise that capitalism, in itself, is a major cause of the situation that the world is facing when it comes to issues such as climate change, global warming, and an increasing propensity of disasters caused by them (Camfield, 2023).

Global capitalism has mostly accepted these tendencies as being a part of the general flow of life, which has made it neglect the larger catastrophic tendencies that lie beneath them (Jones, 2018). The development of capitalism, in reality, has disrupted the natural flow of life which has led to an erasure of the diverse and vibrant histories associated with a particular space (Kiers & Sheldrake, 2021). The erosion of such histories as the previous chapters have shown has caused a general lack of connection among the inhabitants of the urban space, making the possibility of a just transition extremely difficult to be realised. Owing to the seismic effects of capitalism and its extractive methods – be it the middle east or the far east, or be it the United States or the United Kingdom – the idea of a just transition towards more sustainable or climate friendly approaches has become extremely relevant. Just transition is a term that has become widely used to refer to the mode of transitioning towards a different approach towards addressing issues concerning climate change from the point of view of human rights, ecological regeneration, and social justice (Hamouchene & Sandwell, 2023). The politics of just transition represents a phase of human action where human beings actively partake in the process of a seismic shift within how social change is theorised and practised, both as individuals and as a community. It is contingent upon the context within which it is to be applied. A methodological framework that has worked in the Global North or the Global South (in general) cannot be expected to work *in toto* within regions such as the Barak Valley.

The diversity in approaches required has proved to be a difficult terrain for most crude Marxists, as Saito and Goodfellow (2023) highlight, who often analyse social development as a one-dimensional chain of events. The Barak Valley presents a diverse corpus of issues for such politics because it constitutes a highly underdeveloped region, where the politics of development itself has failed to make its mark. Like all other peri-urban regions, Silchar as well conforms to a growing lack of financial autonomy to its local institutions, which has created a certain stagnancy in the steps that the local administration can take in mitigating the effects of floods. For example, the disaster management guidelines of 2024 differ little from those that the DDMAC issued in 2022 and continue to focus excessively on NGOs without much focus on communitarian actions or increasing public accountability.[9] This represents a critical difference that states such as Assam have with other states like Kerala.

Assam and Kerala are one of the few states in India which experience floods on an annual basis fuelled by a combination of climate change and mismanagement by the regional authorities (Outlook, 2022). The major reasons behind the success

---

[9]The reports and other details can be found at https://cachar.gov.in/departments/detail/about-ddma (accessed 07.05.2024).

of Kerala in mitigating the losses of the 2018 floods were based on local governmental structures and the decisions which were taken up after the floods. As Rajamony and Mana (2022) argue, the ways in which the state of Kerala engages with disaster mitigation are intimately dependent on the ways in which it promotes its local governmental bodies and environmentally conscious policies so as to enable more people to 'live with the floods' in a co-operative manner, rather than making it a compulsion. The arguments surrounding 'living with floods' cannot be implemented through a bureaucratic top-down approach but rather require a more intense and people-centric approach. The problems that emerge during a natural disaster affect people not only materially but also psychologically. But it would again be wrong and simplistic to assume that disasters affect everybody *equally*, because contemporary capitalism produces a disproportionate amount of risk for different individuals depending on the socio-biological characteristics that they possess. Local politics, during a disaster, become relevant again because local politics infiltrate into the core of the issues that affect those devastated by a natural disaster. For example, a vegetable seller, stated:

> The municipality could have helped. I went to them only when I first faced the water entering into my home. They said they cannot help anymore because the orders from above had not come until then. By the time, they actually came, my stall and home had been washed away. Had there been a more effective municipality, these problems would not have occurred.

This is precisely the kind of vision that localised variants of egalitarian social justice-oriented theoretical and political tendencies such as municipal socialism present. Municipal socialism, as a tendency, is drawn from and within the argumentative position put forward by the Fabian socialists, who argued that organic changes in the society have to be democratic, gradual, acceptable, constitutional, and peaceful (Webb, 1889/1948). The absence of municipal socialism as an argumentative position under these circumstances is indeed surprising, considering the opposition to neoliberal modes of capitalist accumulation (Leopold & McDonald, 2012, p. 1837) expressed by left-wing and CSOs in contemporary India. The arguments in favour of municipal socialism come at a time when the world has been led to a deathly crisis by the privatisation of services deemed essential for survival within a world that has been devastated by an impending climate crisis and ecological destruction.

With a growing climate crisis, the necessity of local governance has been felt across the world. However, most local governance measures, as Bulkeley and Kern (2006) argue, remain dependent on the higher bodies. This was also noticed in the case of Silchar where most of the local workers working within the essential service sectors remained unable to respond to the floods because they were not empowered enough to do so. The kind of local governance in place does not enable them to do their duties autonomously. Local governance, as Aslam (2007) notes, on the contrary, does not only refer to making laws and rules but 'The objectives of local governance cannot be realised by passing legislations only. The

144 'Natural' Disasters and Everyday Lives

related legislation is necessary, but it is not enough unless its spirit is translated into practice' (Aslam, 2007, p. 51, emphasis self). Neoliberal capitalism makes it very difficult for municipalities to bring in effective policies for sustained self-government through a variety of measures. In sectors such as healthcare and education, the municipalities can play an important role. Because these are local bodies sharing organic connections with the communities, they can have greater impacts on the kinds of situations that disasters such as pandemic create within the communities. In this spirit, the DDMAC (2022) also notes the important role that community volunteers came to play during the floods:

> Assam State Disaster Management Authority in Collaboration with District Disaster Management Authority Cachar trained 318 Community Volunteers under the aegis of National Disaster Management Authority (NDMA). Services of AAPDA MITRA Volunteers were utilised by District Administration in Cachar district in the evacuation, and distribution of relief materials among the marooned people. The volunteers visited the flooded area and relief camp and rendered voluntary services to the Administration. (DDMAC, 2022, p. 28)

During Covid-19, local governance had come up to be one of the major planks of India's response strategy (Deb Roy, 2024a). However, with time, the radical restructuring that had occurred during the course of India's response has waned off. Most of the ULBs find themselves starved off funds and political autonomy, which makes it difficult for them to intervene within the crisis caused by natural disasters. The intervention of local bodies into natural disasters has been noted down by many as being a highly effective method of the governmental step towards disaster mitigation because they can develop more concrete and localised plans if they can counter the various structural, theoretical, and bureaucratic barriers in place (Baker et al., 2012). However, the success has been limited in this respect, as Bulkeley and Broto note:

> The city, so to speak, is now firmly on the climate change map. There is, however, a paradox at the heart of this new-found enthusiasm for the potential for urban responses to climate change. On the one hand, research suggests that the translation of political commitments and policy rhetoric into substantial and programmatic municipal responses has been limited. In essence, municipalities that have pursued a comprehensive, planned approach to climate governance are few and far between and most have encountered significant challenges related to institutional capacity and political economy. (Bulkeley & Broto, 2013, p. 361)

Reforms surrounding the institutionalisation of the local bodies capable of effecting change has always been a major focus of social justice-based politics. They become more important especially within formations that tend to manage

Local Politics, Right to the City, and 'Natural' Disasters   145

rather than obliterate capitalism – the key cause behind ecological damage – which in turn leads to a pragmatic managerial attitude that has aided in the analysis of political issues through a managerial perspective (Panitch & Leys, 1997). These perspectives have included the generation of spectacles that can contribute to a shrouding off of the actual problems that people face during a disaster. The ways in which the DDMAC had to take the help of the community volunteers and NGOs prove the inefficacy of the local bodies to deal with a *flood in a flood-prone area*. A key part of this inability is the financial and political constraints on local administrative bodies. With the rise of finance capital, there has been an erosion of the powers that local institutions hold over the public provisioning of essential services. Finance capital has actively aided the construction of the *super elite* and *the new elites* in the society, who naturally are putting an immense pressure on their respective communities for disproportionate access to services and products. This is because:

> The higher status groups make disproportionate use of local services relative to their need. Contrary to the widespread belief that public expenditure benefits primarily the less well-off there is considerable evidence that many welfare services are distributed in a manner which favour higher social groups. (Stoker, 1991, p. 18).

Stoker's arguments were made considering the provisioning of mostly health, education, housing, and transport and can be witnessed in the case of India. While the elites and middle class, during the 2022 floods, had access to material and emotional resources, it was the marginalised sections of the populace which suffered the most due to the lack of the same. Even people who were parts of the disaster recovery framework did not have access to the basic services at their homes. Municipal workers deployed at the relief camps were often left bereft of the basic protective equipment. These included women from the Integrated Child Development Scheme (ICDS), and the Anganwadi schemes (DDMAC, 2022), many of whom were often denied extra payments or even recognition except some tokenistic mentions. They were, in the words of a worker, used akin to disposable workers who could be called upon when there is a crisis and then 'left alone with no recognition'. The same was the case with those new kinds of service workers – petty-disaster capitalist service providers – who emerged after the floods to mitigate a wide variety of post-flood issues, which include heightened health risks[10] and a growing inflation with many kinds of different services opening up after the floods. In Silchar, these new services included house-cleaning jobs, drain management, dry cleaning, and daily-wage-based chores related to post-flood services within communities, most of which was provided to the middle classes at a premium rate. A house cleaning job was often priced at INR 500 (5 GBP app.) per

---

[10]Encephalitis was one of the major issues to occur in Assam post the 2022 floods. See https://www.thehindu.com/news/national/other-states/encephalitis-deaths-add-to-post-flood-worries-in-assam/article65608737.ece (accessed 14.05.2024).

room – as a daily-wage worker is seen doing in Fig. 24 – while dry cleaning prices were mostly above INR 1,000 (10 GBP app.) for sofas, couches, and bedsheets damaged by the floods (shown in Fig. 25). This was almost a 100% increase from the general rates in the town. An electrical mechanic, a daily-wage worker, and a dry-cleaning service provider, for instance, stated, respectively:

> After the floods, I charged a bit higher than my usual price because my income was reduced to zero during the floods. There were like fifty complaints of damaged electrical equipment. So, I also charged a bit more. It was to compensate for the lost income during the floods.

> After floods, I went to work at many houses. Most of them did not bargain when I charged a high fee, but some of them did, and I also did not object. Some people might have some financial issues, many things were destroyed during the floods. But I got work after the floods, so much that I had to get another person to share the work. Both of us earned.

> The business just boomed after the floods. Many people came with bedsheets and sofas. Some we took, some we did not take. The ones that we could not take ended up on the road.

Fig. 24. A Daily-Wage Worker in Action After the Floods. *Photo credit*: Author.

Fig. 25. Sofas and Mattresses Destroyed By Floods, Some of Which Were Handed Over to Dry Cleaners. *Photo credit*: Author.

Municipalities can play an important role in the expansion of the network of amenities, such as piped gas, health facilities, education, and drinkable water, which are still hard to reach in most Indian cities (Aiyar, 2020; Singhal & Mathur, 2021) and more so during a crisis such as a flood. Without the help of the community, any municipality-based political establishment will find it extremely difficult to implement reforms that engage directly with the community (Towler, 1909). Local governance or the control over resources by municipalities is part of the long heritage of marginalised class politics focused on the assertion of their rights within the urban space engaging with everyday acts of resistance (Stromquist, 2023). The extension of local self-governance is an issue that is intimately associated with the ways in which individuals function because local governance, mostly at the municipal level, is associated with the immediate community that the individuals are a part of. However, communities operating under capitalism are not free from the impact of capital's sustained logic because of the totalising nature of capitalism today (Anderson, 2020; Lebowitz, 2020). When more and more individuals from a certain community start getting upwardly mobile in large quantities, there also occurs a proportional growth of the community's own internally bred elite, which then carry the logic of capital with them back to the community. This can be seen in the ways in which various middle-class individuals

148 *'Natural' Disasters and Everyday Lives*

acted during the floods in Silchar as well exploiting the connections that they have with the socio-economic and political authorities in place.

Urban spaces are often the representations 'with people and their activities as an ongoing set of possibilities' (Knowles, 2003, p. 80). These are sites where the activities of an individual are related to:

> [...] inherent capacity of [the individual] to direct behaviours of the person intelligently, and thus functions as a special kind of subject which expresses itself in a pre-conscious way ... usually described by words such as 'automatic', 'habitual', 'involuntary' and 'mechanical' (Seamon, 1979, p. 41)

critical to the formation of one's lifeworld. Similar to the other activities, the daily 'involuntary' actions are also a result of these processes which intersect with how individuals think and perceive the space around them. These practices which engage with acts of directly voluntary actions such as walking and chatting instil within the individual a sense of belonging to the place instigated by the perception of dwelling in those places (Rodaway, 1994; Seamon & Mugerauer, 1989). The 2022 floods of Silchar allowed some of the people of Silchar to reinvigorate this sense of dwelling by re-establishing their connection to their communities. This re-establishment was often through the unproductive activities that had withered away under the pressures of their working lives. It happened mostly through a re-imagination of the local sense of place that most of the individuals had relegated to their subconscious selves. The contemporary society is one where, as Kumar-Rao (2023) mentions, local histories and customs are increasingly at the risk of fading into oblivion. The marginalisation of local histories disrupts the sense of a community, which in turn affects the psychological security of individuals. Feelings of being secured ontologically within an urban space are contingent upon how satisfied the users are with the 'immediate, everyday environment, specifically the level to which people perceive their living environment as satisfying their needs' (Jabareen et al., 2017, p. 2). This sense of security can strengthen the sense of a community along with the attachment that an individual feels towards the space and the place, the feeling of cohesion, and the overall social life within a space.

Individuals do not experience urban reality as they want, but rather they experience it as *it is*. As Lefebvre remarks, 'Every space is already in place before the appearance in it of actors; these actors are collective as well as individual subjects inasmuch as the individuals are always members of groups or classes seeking to appropriate the space in question' (Lefebvre, 1991a, p. 57). The development of urban spaces is also impacted by cultural preferences of the various individuals and groups therein, who form distinct communities based on common institutional figurations (Ley, 2004, p. 153). During the 2022 floods, a certain section of the populace enjoyed a growing communitarian lifestyle, as a middle-aged man states:

> I have been living in this neighbourhood all my life, but in the last 10-12 years had lost connection with all my friends, mostly

because most of them have become rich and well-off but I have remained there. But when the floods came, I realised that I had no other option but to seek them for help.

Such desires for a co-operative and non-exploitative community, as Bauman (2001a) notes, stems from the existing deep insecurities in the everyday lives. The flood victims represented this transformation, in their appreciation for those who exhibited for them a growing sense of a community and extended a helping hand to them despite the differences that they possess. Such factors contribute to a reduction of the ontological insecurity, more so during times of a natural disaster in a region with high rates of multidimensional poverty, slow municipal responses to post-flood issues – such as the post-floods waste, as shown in Fig. 26 – and heightened fault lines in a disaster-struck region.[11]

Floods lay explicit various kinds of marginalities that a society faces, not only with regard to its own members but also in terms of the relationship that the

Fig. 26. Post-Flood Waste in Silchar. *Photo credit*: Author.

---

[11] See https://www.barakbulletin.com/en_US/into-the-twilight-imagining-a-future-for-barak-valley-under-the-spectre-of-poverty-hegemony-and-sedition/ and https://scroll.in/article/1028061/how-communal-rumours-hid-the-truth-about-the-deluge-in-assams-silchar (accessed 06.05.2024).

150 'Natural' Disasters and Everyday Lives

members share with the broader structures in place. Floods, as this book shows, do not only harm the material resources that people and communities possess but also affect them at a deep subconscious level that changes their perception of reality, as can be seen in the case of the Dalits and Muslims of Silchar. Perhaps, the most important example that can be evoked in this regard is the kind of feelings that people attach to their houses. Houses that people build for themselves and their families are different from mere buildings because they 'project onto the land a conception of the world' (Lefebvre, 2003, p. 22). Community formation processes are intimately related to the factors of trust and recognition and are manifestations of networks of support and care essential for an individual (Alexander et al., 2007, p. 793). Alleyne (2002) and Putnam (2000) have argued that the classical idea of a 'community' usually revolved around material factors like home and land, and floods affect both these factors. Under disastrous circumstances, community formation processes transcend such crude objective analysis focused on commodity possession. The loss of one's home *as one knows it* still reverberates among individuals even if the premodern ideas of community have given way to the establishment of looser and more individualised networks of association owing to atomisation and alienation.

Cities are spaces within which a diverse group of people, often abject strangers to each other, come together to form a common cultural norm, even if they are reluctant to do so (Harvey, 2012, p. 67). Cities reflect the most successful human attempt at remaking the world in the image of the actually living individual's thoughts and desire. However, 'if the city is the world which [the individual] created, it is the world in which [the individual] is henceforth condemned to live' (Park, 1967, p. 3). Cities reflect the inhabitants and users therein as 'what kind of city we want cannot be divorced from ... what kind of people we want to be' (Harvey, 2012, p. 4). The norms and rites, which are part of the urban civility, are also a part of the monumental effect of institutions which frame urban norms around the space they create and inhabit. In these circumstances, acts of community formation are bound to happen following a pattern of rules. It is then the task of the urban practitioners themselves, the people who 'live and practice the urban' to transform those communities into instruments for accessing the *Right to the City*.

Chapter 6

Towards a Humanist Politics of Climate Change

The human society globally has not been able to manage the delicate relationship that it shares with nature, which has led to increasing climate-related disasters. The word 'disaster' stems from Latin where it means '"ill-starred," literally a malevolent omen from the heavens. But climate-change related disasters are no longer a matter of bad luck. We have tilted the odds toward catastrophe, particularly in the places that did the least to cause the problem' (Holthaus, 2020, 'A Living Emergency'). The devastation that was caused by the floods is a testimony to this argument. It is no surprise that natural disasters have grave consequences for the economic fabric of a society (Bitzen et al., 2019), which have continued to become more severe over the years in neoliberal societies because of the growing inequalities and inability of neoliberalism to cater to overall well-being of the society (Ostry et al., 2016). Policies focused on countering climate change need to be enacted locally but have implications that are international in nature (Bulkeley & Kern, 2006). These policies need to be devised in such a way that they do not only take the local contradictions into account but also consider the dialectical relationship that they share with global production systems in place. This conforms to a total analysis of capitalism, without creating a fetish out of one's local circumstances.

Ecological destruction and its associated climate change is a part and parcel of the capitalist development process and has been one of the mainstays of the contemporary society as one knows that it produces an increasing plethora of issues pertaining to sustainability and global warming, steering the human society towards a 'war'-like situation between human society and nature (Malm, 2016). Jason Moore (2015) argues that capitalism today is an organisation of nature itself. Foster (1999, 2000) argues that there has occurred a metabolic rift between human society and nature. Natural disasters such as floods have become more frequent in Assam with more than a million people being affected by them annually.[1] The Intergovernmental Panel on Climate Change (IPCC), formed in

---

[1] See https://www.thequint.com/opinion/why-have-floods-in-assam-become-an-annual-scourge#read-more (accessed 05.05.2024).

'Natural' Disasters and Everyday Lives:
Floods, Climate Justice and Marginalisation in India, 151–163
Copyright © 2024 by Suddhabrata Deb Roy
Published under exclusive licence by Emerald Publishing Limited
doi:10.1108/978-1-83797-853-320241007

152 'Natural' Disasters and Everyday Lives

1988, has repeatedly highlighted the possibility of impending climate catastrophe, one which has constantly been neglected by those in power. As Park argues:

> The implicit assumption built into the logic of capitalism is that economic growth is directly correlated with social prosperity. As the economic system expands exponentially, society will consequentially improve at an exponential rate, for society is theoretically better off for having more wealth than not, irrespective of how that wealth is allocated. The finite nature of resources is not a concern because when any given resource becomes scarce, the market will naturally produce an alternative. This is founded on the idea that if there is a great enough demand for a product, price signals will mobilize firms to meet that demand. (Park, 2015, pp. 191–192)

Capitalist models of urbanisation are based on the persistence of uneven development and a constantly degrading environment (Greenhalgh, 2005; Peck et al., 2009). Small towns and peri-urban regions within such a social setting remain important actors of economic and social growth and the management of natural resources because of their abilities to serve as points of transit within the regional, national, and global production mechanisms (Tacoli, 2017; Tiwari, 2019). Despite these characteristics, the impacts of urbanisation and climate change on small towns still remain grossly under researched (Steinführer et al., 2016). The development of small towns and peri-urban regions in India, however, has always been fraught with numerous contradictions, the most critical of which is the general lack of accessible services and a constant fetishisation of biased urban development that mostly overlooks these regions (Guin, 2019). Along with this, small towns have also become vulnerable to a lack of social capital[2] among its residents that results in lower civic engagement while simultaneously exhibiting a growth of market capitalism and petty production units (Besser, 2009). The overall nature of capitalist urbanisation is such that productivity has become a central point for being a part of the urban space, which creates in the minds of those in hyper-productive jobs a sense of guilt and a fear of missing out – popularly known as FOMO in the contemporary times – which makes them more frustrated than the situation actually warrants. Local administrative networks, in situations such as this, can be helpful because these networks are ingrained within the community spaces, which are more intimately associated with the individual. While abstract universal bodies such as the state cannot afford to be in contact with each individual, the local community-based institutions can do that under ideal conditions. However, under neoliberalism, most of the local administrative

---

[2]Social capital in this context refers to the definition given by Putnam (2000) and 'is defined as the relationships between people characterized by trust and norms of reciprocity that can be utilized for individual and collective goal achievement' (Besser, 2009, p. 185).

bodies have found it difficult to assert themselves amid the growing impact and influence of the neoliberal market.

To some, neoliberalism has in fact caused a resurgence of the idea of India as a superpower because of 'its sustained democracy, its embrace of technology – especially informational technology – and its open, plural and diverse society' (Pande, 2020, p. xi). While the first one is true – to a certain extent – it is the second one that has become the major point of contention under the far-right regime that governs contemporary India. The manifestations of the far-right governmentality were also visible during the floods of Silchar when political connections were exploited by the middle class in garnering better benefits from the authorities, as this book describes in detail. Disasters make it explicitly visible the political capital that any community or class possesses, which, in turn, often come to determine the chances of survival that a particular individual or community possesses. Disasters affect and get affected by the long-term choices that people and societies make. There has been a growing body of thought that has advocated that natural disasters *can be made to* act as portals for social change amid the 'climate chaos' that capitalism creates. And urban spaces, as Dawson (2017) argues, are one of the chief drivers (as well as the prime victims) of the same. Contemporary cities are stuck right at the centre of the growing implications of climate change with growing industrialisation and global warming (Narain, 2017; Ramesh, 2019), becoming increasingly prone to disasters. Natural disasters – or catastrophic events – provide the society with certain opportunities, which if used in a proper democratic fashion can prove beneficial to an egalitarian vision (Gittlitz, 2020). Most of these tendencies see in a disaster a possibility of hope for a more democratic and radical future based on disaster communism which can be defined as follows:

> Disaster communism is not divorced from existing struggles. Rather, it emphasizes the revolutionary process of developing our collective capacity to endure and flourish: a movement within, against, and beyond this ongoing capitalist disaster. [...] 'Disaster communism' adds a clarifying epithet to the already long-ongoing political project that pits itself against the state and capital and overflows their bounds. It orients the movement of a collective power that, while rendered palpable during extraordinary disasters, was there all along, especially in places and among groups who have been experiencing the situation of ordinary disaster for hundreds of years. Climate change throws the skills central to those struggles into stark relief. (Out of the Woods, 2018, para 5)

Disaster communism sees in disasters a window to rupture the capitalist society from within, creating newer forms of community that go beyond capitalist forms of alienation and atomisation. Heron and Dean (2022) argue that the solution of the impending climate catastrophe can be based on nationalisation and further regulation by the state. To do this, it has often been prescribed that a new form of party is required (Heron, 2013), one which is distinct from the one that is

154 'Natural' Disasters and Everyday Lives

commonly visible in contemporary anti-capitalist politics. However, these visions often work with the assumption of a distinction between the people and the party (Dean, 2018). The anti-capitalism of such left-wing politics often reduces itself to an alternate form of statism that does not critique the fundamental contradictions in place. For example, speaking about the Covid-19 pandemic, Heron and Dean note:

> Few are persuaded by the denial of the political nature of climate change [one of the major precursors to disaster capitalist tendencies]. Persistent mobilisation by the grassroots activists has placed climate clearly on the political agenda ... voters recognize climate change as a matter of politics: it's an issue that simultaneously divides and necessitates a political response. Moreover, as is clear to nearly everyone, the scale of the catastrophe requires a state response. (Dean & Heron, 2020, para 21)

The state here is sought to be a welfarist and democratic institution that reflects the Keynesian welfare state of the yesteryears. However, as Parenti (2016) argues, the state in recent times has increasingly becoming aligned with the Anthropocene, exerting a biopolitical control over the people. Visions of apocalyptic communism have mostly focused on the imagination of an anti-capitalist futurism based on automation and a communism of luxury (Bastani, 2018) or in the vision of a revolutionary state acting as an agent of transition between capitalist state and the workers' state, altering property structures, operational modes, and formulating relations of production that are free from the desires of profit accumulation (Posadas, 1969). Posadas had argued:

> [T]he leaders who create *Revolutionary States* originate from the capitalist regime. They come from bourgeois organisations, bourgeois institutions, the army. The army often takes a leading role in the revolution because it is often the only constituted power. This is how the *Revolutionary State* comes about. (Posadas, 1969, p. 14)

The relevance of such arguments in peri-urban India remains fragmented because these areas do not reflect the ideal state of development that is assumed within most theoretical traditions. In most cases, the state and the civil society act in unison during a natural disaster, as Wilikilagi (2010) shows, which make the amalgamated formation an ideal entity in sustaining (and creating newer forms of) a corporatist and de-humanised form or urban life (Warner, 1983). This is because despite most of the scientific and technological progress that capitalism has made in the previous decades, it still remains a key cause behind the crisis that humanity faces today with the global wave of climate change (Malm, 2020). The ecologically insensitive development that capitalism produces has become one of the major drivers of climate-related disasters in the contemporary society (Kolbert, 2006). The growing biopolitical domination of capitalism has given rise to tendencies such as disaster capitalism to emerge even in the farthest hinterlands of the world (Klein, 2007; Lowenstein, 2015). Mainstream disaster capitalism basically refers to making profits out of greed, but the most visible kind of disaster capitalism that one could witness in Silchar was petty-disaster capitalism, where the marginalised took it upon themselves to exploit the hitherto exploiters or the privileged. The transformation of the exploited into the exploiters provides one with a manifestation of the changes that a disaster brings in within the society and exhibits the possible consequences of climatic disasters at a broader level. Climate change, global warming, and their associated effects such as harsh droughts and increased floods in general often have great implications for the society in general, giving rise to tendencies which are destabilising for the global economic structure (Adam & Tsavou, 2022; Ramesh, 2019). The solutions to such issues, however, cannot simply be based on technological development or philanthropical welfarism of the state.

The INR 3,800 (4 GBP app.) granted to each family devastated by the floods by the Assam government satisfies merely a part of the marginalisation that they faced during the floods. Such philanthropic approaches, as Malm (2016) shows, merely become another cog in the giant wheel of capitalist domination. At the same time, mere scientific advancements as well will not solve the broader issues at play. Das and Mitra (2003) have elaborated the scientific and technological steps – such as strengthening embankments, construction of more sluice gates, and more watershed management practices – taken up by the successive governments in Assam. Cook (2019), while discussing about such steps within a flood-prone region, has argued that such types of developmental politics and steps often occur without much acknowledgement of the ecological impacts that they create. And scientific development without human or ecological considerations are often bound to fail as Lochbaum et al. (2015) have argued.[3] The solution to climate change and its associated issues such as annual floods cannot be solved only through scientific interventions but requires deeper social analysis that go beyond the contours of the generally advanced ones by most bourgeois and capitalist formations.

Disasters always bring forward certain important ideological and material contradictions to the foreground of analysis contingent upon the context of the disaster. A radical vision towards climate justice amid such situations needs to take into cognisance that the relation between human beings and nature is one which is increasingly being converted to a master-slave one (Foster, 2009). It is dictated by a growing regime of private property-based ownership model where the exploitation of nature has become synonymous with development (Kumar-Rao, 2023). For the people of Silchar, the floods exhibited the kind of risks that they harbour within their everyday lives. For most of the people from the marginalised sections, the major issues were not the water in their homes, but the extreme

---

[3]It is important to mention here that in terms of ecological innovation, India still lags far behind its Western counterparts so much so that Mridula Ramesh (2019) argues that it is unlikely that – with the scientific educational level in India – a permanent solution to climate change would emerge from India given the current state of scientific and technological education in India.

156 'Natural' Disasters and Everyday Lives

poverty that characterised their lives. The key to resolving such issues occurring due to climate change under contemporary capitalism remains within the solutions that one can envisage towards solving the human issues that marginalised individuals face, which may or may not be directly related to climate change. In other words, there is a need to address the totality of human existence because capitalism today works as an organic system (Lebowitz, 2020) affecting the total human existence.

In recent times, human society has made great progress in technological advancements concerning to disaster prediction and mitigation (Krichen et al., 2024), the reflections of which can be seen in the local disaster management plans put up by the district administration even in the case of Silchar as Chapter 5 highlights. However, such managerial techniques do not question the degrading communitarian life in the city – one that had been exhibited during the 2022 floods, with most of the middle class actively manifesting a kind of class hatred and hostility towards the extremely vulnerable that presented a highly disorienting and anti-solidarity view of the society. One of the founding figures of sociology, Emile Durkheim (1893/1933, p. 79), had theorised the social system as having a life of its own producing collections of socially acceptable 'beliefs and sentiments common to average citizens of the same society' and termed it to be 'the collective or common conscience'.

This is because during any natural disaster, the expectation that people have from others changes drastically. The middle class expect the political class to help them, while it itself proves to be extremely hostile towards the marginalised and the vulnerable. The 'meta power' that individuals generate under neoliberalism makes them structure different kinds of relational power structures within their immediate surroundings. Most of such power is held by individuals from the dominant community in place that makes it difficult for the marginalised to engage with the politics of just transition. A key part of this is the evocation of performativity and reciprocity because 'when an individual appears before others he will have many motives for trying to control the impression they receive of the situation' (Goffman, 1956, p. 8). It must be taken into consideration that even during a natural disaster, there are certain aspects of performativity that continue to play a role within the society because contemporary capitalism has transformed itself into a form of realism (Fisher, 2009). Heron (2013) argues that in order to break free from this problem of transition, an autonomous party form is required. However, the basic problem in this conceptualisation is that any kind of party form because of its adherence to established norms often is destined to recreate similar contradictions that it had initially set to oppose if its basis does not rest upon a revolutionary philosophy (Dunayevskaya, 1973).

This is also crucial to the sustenance of a capitalist urban space which strives to ensure a certain *guaranteed level of competence and a specific level of performance*' (Lefebvre, 1991a, p. 33) by using instruments of power. Monuments are the most important instruments in the larger arsenal of economic and political instruments. By normalising a certain culture, monuments enable the reproduction of ideological formulations vital to the sustenance of the conceived urban space and culture that mostly acts against the interests of the marginalised and vulnerable (Merrifield, 2014). The social totality, the level at which important decisions are made, cannot be separated from the local dynamics produced by these processes, the level at which those decisions are perceived and lived through everyday actions (Lefebvre, 2008). The dialectical relationship between the total and the local produces the community – an agent attributed with the function of rediscovering the already existing relationship between the individual and the social. These processes are becoming increasingly difficult because under neo-liberalism, an increasing number of individuals are suffering from lack of basic necessities including but not restricted to housing (Aguirre et al., 2006; Holm, 2006), which could be seen in the case of the 2022 floods. One of the key reasons for the same is the political irrelevance of issues such as housing and public distribution in India, especially in regions such as Assam which have continued to be seen as regions entrapped within developmental issues.[4]

Unlike other countries such as China and the United Kingdom where public housing is a major political agenda, India does not have any such policy frameworks within its socio-legal and political infrastructures. The various schemes initiated for the same also affect a very miniscule portion of the populace. The bureaucratic failures of the Indian state and the lack of adequate focus on the same by the liberal media have contributed to the marginalisation that the region faces. While the numerous accounts by the journalists on the floods of Barak Valley have proved to be valuable insights to the scale of the devastation that the floods have caused, most of them, except Krishnamurthy's, have failed to go beyond the 'journalistic flair' that liberal media is often found fetishised with. A key reason for the same is the marginalisation of local journalism under neoliberalism (Sullivan, 2020). Most of the journalistic accounts that have come from the region, especially those during the floods, did not address the basic questions – or rather addressed them insignificantly – concerning the class, caste, and gendered nature of climate change and the effects that natural disasters have in the everyday lives of the people. They fail to cut through the actual ideological ramifications that natural disasters bring forward. It is important to realise that most of the journalists work for media houses who have corporate interests and as such often remain bound by structures that create the enabling conditions for neoliberal capitalism. This becomes more complicated because the state and community in India do not operate with a distinct segregation between them.

The basic issue with the liberal media remains its fetishism for parliamentary politics and its propensity to succumb to liberal political trajectories. This is true for reportage on most natural disasters. For example, Anu Lal writes regarding the 2018 Kerala floods: 'As a conclusion, the point that I want you to understand is that you should try to stay alive. Staying alive is more important. If you are alive, you can make money, help others, and save your family too' (Lal, 2019, p. 29). These statements do not consider the core contradiction that exists in

---

[4]For more details on underdevelopment in Assam, see https://timesofindia.india-times.com/city/guwahati/assam-among-10-least-developed-states-panel/article-show/23117187.cms (accessed 08.05.2024).

158 'Natural' Disasters and Everyday Lives

contemporary society, i.e. the one between capital and labour, which creates the conditions of exploitation. Within such an unequal social structure, the struggle for survival takes place on a highly unequal terrain. The middle class has a better chance of surviving natural disasters than the marginalised, as the 2022 floods have also shown. This is because various 'lifesaving' means such as social media and the usage of the same for voicing their concerns remain a privilege that is reserved for the few. Hence, despite the fact that many of the middle-class people had to – as a middle-class individual mentions – 'climb up the top of the water tank of [their] terrace' to get a phone connection, they still remain much more capable of mitigating the effects of the disaster because of the social capital that they come to possess which makes them capable of, as a flood victim put it, 'knowing the right person to call, while [the poor] people mostly call their own folks who themselves are struggling for survival'.

The generation of an ecologically just society needs to first base itself upon the radical possibilities that Marx's (1844/1973) 'new humanism' proposes to bring in a vision where a human being is within a dialectical relationship with nature. The new human becomes capable of addressing the totalitarian nature of capitalist exploitation that develops organically taking different facets of human life, including the state and the party, into account (Lebowitz, 2020; Tronti, 1962). Lefebvre (2002, p. 33) described the 'total human phenomenon' as being a dialectical whole formed out of the interaction of needs, work (labour), and pleasure. Egalitarian modes of local administration can prove to be a great advancement in this regard, which has been a key concern for most Marxists. Left-wing policies have mostly advocated for greater decentralisation of the process of governance in India, with further ability to control provisioning and distribution being vested upon local bodies (Deb Roy, 2024a).[5] However, most mainstream left-wing tendencies have often worked with the assumption that the idea of 'people' can be substituted by 'decentralised representative governance'. This has often resulted in a very crude version of radical politics (Bookchin, 2015, 2021).

The relationship between individuals in societies such as India mostly gets dictated by the structures of the community in place. The community dictates the division of labour in place in much of the Indian society. Be it within the sphere of waged labour or within the sphere of unwaged labour – performed mainly by women – or

---

[5]As Taylor-Gooby argues, 'Most left strategies include higher public spending and more equal social provisions, but public opinion rejects both tax rises and greater generosity to the poor of working age' (2015, p. 126). The right-wing on the other hand, focuses more on 'private enterprise-led recovery, work ethic values and policies that exclude less deserving groups' (Taylor-Gooby, 2015, p. 126). The strategies that the left has adopted in most countries have been in accordance with these principles. The socialist left's advocacy for more public ownership of essential services has been one of the mainstays of left-wing economic arguments. Taylor-Gooby's argument points towards a trilemma, that the left faces when designing effective public policies. The trilemma is constructed by the need to respond to growing global economic crisis competitively at par with other forces, address popular public opinion questions, and the development of policies that are simultaneously income-generating and inclusive.

unrecognised undignified labour, mostly performed by the minority and other marginalised communities. The relationship between individuals becomes the most critical aspect of the vision of climate justice, one which does not feature within the alternative statism frameworks. This is because any form of statism does not counter the very basis of capitalist exploitation, the existence of private property. That is the key to understanding exploitation under capitalism, as Marx (1844/1973) had pointed out, one which cannot be grasped by any form of 'revolutionary state' or 'mainstream party formulations'. Visions of a radical state or party form remain highly contradictory because the state will eventually come to be associated with forms of a completely subsumed individuality contradicting the basic idea of Marx (1973) who saw individuals and social structures in a dialectical relationship with each other.

The kind of revolutionary states that Posadas argues in favour of, or the regulatory state that Heron and Dean advocate for, embed within themselves a subsumption of the people to a certain extent. They often do not acknowledge that the self-development, self-activity, and self-movement of the masses are critical to the revolutionary conception of an anti-capitalist movement (Dunayevskaya, 1973), even within the contours of an ecosocialist movement. The key to developing a radical ecological movement is to transcend the model based on private ownership which creates the basic capital–labour contradiction in the society, which also extends to the environment because nature is also mostly privately owned by Big Capital today (Foster, 2009), the implications of which are found in the ways in which dams and dikes are frequently used by the state in furthering their developmental agenda (Nagendra & Mundoli, 2023). Flood control does not only mean the construction of reservoirs and dams but also means taking note of the social problems which are caused by floods (Bhattacharya, 2003). The history of flood control in Assam is a long one as is exhibited by the research conducted by D'Souza (2016) and Baruah (2023), but the contemporary situation that the state faces as a whole is one which continues to be lacking in infrastructural support and a vision of alternative developmental politics. In terms of ecological justice and climate change, relating the floods of Silchar to broader political frameworks, it is important to understand, as Singh (2009) has right noted, that merely focusing on local issues would not provide a fruitful or comprehensive answer to the contemporary woes of the marginalised in the South of the Global South (SGS). Mainstream models of localism, such as municipal socialism, remain inept at understanding the complete spectrum of human nature and individuality and thus fail to contribute to a more radical understanding of climate justice and disasters. The activities that people engaged in such as fishing and chatting with others about one's conditions provide a window to a possible version of post-capitalist scenarios where human existence is not associated with hyper-productivity and one's ability to produce surplus value.

The social inequalities that characterise contemporary capitalism restricts access to means of social mobility which in turn affects the conditions that people face during disasters. The most visible aspect of this remains the agrarian and environmental degradation that capitalism produces that leads to malnourishment, poor health, falling life expectancy, and other socio-economic issues. The major challenge in this context for most political and social movements remains

160 *'Natural' Disasters and Everyday Lives*

the establishment of participatory democracy and universal access to the basic needs of sustenance (Brownhill, 2009). Communities and local governance can prove to be a critical part of this vision. When communities are considered to be networks of ideas and people rather than static assemblages, they transform into living entities with a life of their own. The people of Silchar showed glimpses of such a communitarian form during the floods when they engaged with activities.

The government's many plans for urban restructuring following a major disaster such as a flood or an earthquake cannot accomplish much unless they satisfy the urban needs of the most marginalized, and they will not be able to become fruitful and effective urban policies in terms of accessibility and inclusion in major urban programmes, policies, and reforms. In most cases, the individual violence that middle-class people exert merely forms a part of the broader systemic inequality in place. Along with the individual's own motives, there is the additional control exerted by the monumental institution in place, which tends to direct it to a set of specific responses. In other words, the prescribed intention remains to exert the notions of pre-existing set of rules regarding how individuals are perceived by others as well as by the selves, both vital to mechanisms of social interaction while the actual feelings of the individuals can only be ascertained through their involuntary responses (Goffman, 1956; Scott, 2009) towards the events of the communities and the spaces they inhabit. There are aspects of social control, stigma, and the desire to 'be a part of a culturally sanctioned advanced community' which play a crucial part in these processes. The hidden nature of the actual feelings and emotions enables the individuals to effectively navigate through their lives in the community.

In general, communities are formed through 'positive and sentimental attachment to neighbours, local establishments and local traditions' (Suttles, 1973, p. 35) when people in social relationships generate mutual concern for each other relying on their emotional connections and patterns of interpersonal interactions (Fine, 2021). Most of these interactions in their ideal states evoke a sense of social capital as Putnam (2000) had described. During the 2022 floods, such tendencies of community formation became largely contingent in accordance with the class that one belongs to. The middle classes could reconnect to their communities and roots albeit within difficult circumstances, creating a sense of communitarianism. On the contrary, the marginalised and vulnerable classes found themselves at the mercy of their fates, having to search for avenues to survive the effects of a disaster. This led to the commercialisation of essential goods during disasters, which have been often performed by working individuals themselves – some out of need and some out of greed – creating the ideal grounds for the philosophy of the market and disaster capitalism to thrive. Disasters thus change the experience of the nature of time and space for individuals. Berman (1983) mentions the experience of time and space with respect to both the self and the others to be a vital criterion of living through modernity itself. Routines enable the proper functioning of solitary activities, which become important elements to construct the consciousness of the individual.

Consciousness forged out of a feeling of being secured enables them to navigate their ways through the everyday life created by and creating the urban space around themselves. It is this consciousness which enables the individuals to live

Humanist Politics of Climate Change   *161*

and work with others while protecting their ideas of their own selves within the urban space. When people live in close proximity with each other – that is usually the case during a natural disaster that displaces people – this balance gets disrupted, and people turn to following new mores and norms. These newer forms of communities though revolutionary and prove to be windows of hope, as Out of the Woods (2018) argue, but as the 2022 disaster in Assam shows, they immediately revert back to their original capitalist form once the society regains 'normalcy'. This is because these newer forms of community do not challenge the fundamental reification of human relationships that form the core of the capitalist commodified social existence – the rupturing of which is critical to any form of humanist alternative (Hudis, 2021).

This process takes place within the everyday lives of individuals eroding the very basis of traditional forms of social security that an individual enjoys. Social security alters the ways in which individuals react to the society and the kind of aspirations that they have which, in turn, also alters the ways in which the marginalised and vulnerable individuals come to define concepts such as poverty, unemployment, and precarity in their everyday lives. These are often, as this book describes, very different from the middle classes. The middle class is often accredited to be the driving force behind the democratisation of the society. However, because of the class privileges that the middle class possesses, it is often cut off from the stark reality India faces on an everyday basis. This makes the middle class harbour a very restricted view of the society and the implications of climate change for the broader society. The associations that middle-class individuals form – the resident welfare associations for example – had become one of the key players behind the obliteration of the right that the marginalised people have to the urban space. These associations, during the floods, have been the major organisations behind stopping the displaced people from entering the buildings when they could not find spaces in the relief camps, as this book has narrated. Most of these processes hindered the vulnerable people in forming a community of their own, one that did not replicate the exploitative conditions that they found within their own surroundings.

Communities and their dynamics take an active part in the formation of these spaces. Within most urban spaces, the individuality of the inhabitants develops through their interactions with the existing social diversity and everyday patterns of life (Watkins, 2005). During the floods, the everyday patterns of life are usually dictated by a sense of hostility and class hatred, which the middle class and the elite had been found to exert upon the marginalised and the vulnerable through their actions. This is a manifestation of the petty-bourgeois and anti-working-class role that these associations of middle-class individuals have come to play in contemporary India (Kamath & Vijayabaskar, 2009). This gets aptly reflected in the ways in which people articulated the reasons behind the lack of electricity during the floods. On a general day, the lack of electricity would be blamed on the state in Silchar,[6] mostly along the lines of the argument that the state is not doing enough to provide sufficient electricity to the people. However, during the

---

[6]Most of the electricity distribution in India continues to remain under state control.

162   'Natural' Disasters and Everyday Lives

2022 floods, the focus was not only on blaming the state but also upon a constant appreciation of possible privatisation. The economic changes associated with neoliberalism play a large role in rendering the public sector relatively power-less which makes them unable to effectively intervene in the society even during natural disasters.

Tendencies such as Posadism or crude forms of disaster communism lay hopes on higher bodies because, to them, as Gittlitz (2018) argues, any form of higher beings would have solved the contradictions that common human beings face and as such would be able to help human societies in resolving the contradictions that it faces. In doing so, however, these tendencies do not take into account the fact that all kinds of 'higher beings' would also – in all probability – function as monuments themselves incapable of understanding and analysing the internal contradictions in place. The movement against the state as a force of regulation and authority, however, has been a two-pronged one, one that has simultaneously been resisted both by the market – with a view to replacing the state – and by the long-drawn autonomous movement of the marginalised. However, the most critical point to realise at this juncture is that even though the latter movements typically have characterised the state to be a force that needs to be there to ensure social justice, they have not been completely against the state. These movements have often been in support of the state being restored to a welfare-oriented state, where the needs of the marginalised and the poor take precedence over those of the ruling class. More often than not, such arguments have led to a kind of statism within the overall discourse of radical social change. Such a character of some social movements makes them a part of the overarching capitalist social reality where they have resulted in being only able to negate the superficial aspects of capitalist degradation of the environment and the inequality that is associated with it. This vision needs to focus on the dialectical relationship that individuals share with society and the state as had been highlighted by Marx (1973) and how it affects the marginalised people during times of distress.

Ecological crisis is a reality of contemporary capitalism, one from which it is difficult to shield oneself from, especially if one is from the marginalised popu-lace. The technological development – often based upon the premises of benefit-ting everyone – has often been used to benefit a minor section of the population (Deb Roy, 2024a). A radical agenda for ecological justice needs to emphasise that a just transition within an urban settlement cannot be exercised until the most marginalised sections of the population can claim their right to the space that they inhabit. Their access to the space, in turn, depends upon the kind of urbani-sation that takes place within a particular space. Just transition towards a more climate justice-oriented framework thus cannot be envisaged unless the marginal-ised can exercise a right to the city – a right which does not only give them a legal right over the urban space but also an ideological right over the city.

The climate justice that most activists speak of cannot be brought about merely through activism focused on climate change but rather have to be focused on the deep-rooted inequality that characterises contemporary societies. This requires a dialectical unison of activism and serious analysis. The struggle for climate justice in these regions – and across the globe – does not only require a local solution but also a global one that conforms to the global struggle against not only climate change but all forms of inequality in the society based on the idea of the *new human* of Karl Marx who does not pose as an adversary of nature but rather as a being who is one with nature. To achieve that, forms of higher powers need to be relegated in favour of human autonomy free from social control of higher powers that tend to dehumanise human existence and reduce the same to quantitative samples or passive victims. The actual self-movement of the masses reinforcing their own autonomy – against capitalist exploitation, uneven development, and alienation – needs to be at front and centre of the struggle.

References

Acevedo, S., & Novta, N. (2017). *Climate change will bring more frequent natural disasters & weigh on economic growth*. IMF blogs, November 16. Retrieved May 6, 2024, from https://www.imf.org/en/Blogs/Articles/2017/11/16/climate-change-will-bring-more-frequent-natural-disasters-weigh-on-economic-growth

Adam, A., & Tsavou, E. (2022). Do natural disasters fuel terrorism? The role of state capacity. *Economic Modelling, 115*, 105950.

Aguirre, A., Eick, V., & Reese, E. (2006). Introduction: Neoliberal globalization, urban privatization, and resistance. *Social Justice, 33*(3 (105)), 1–5.

Ahluwalia, I. J. (2019). Urban governance in India. *Journal of Urban Affairs, 41*(1), 83–102.

Ahluwalia, M. S. (2002). Economic reforms in India since 1991: Has gradualism worked? *Journal of Economic Perspectives, 16*(3), 67–88.

Aijaz, R. (2007). *Challenges for urban local governments in India*. Asia Research Centre Working Paper 19. London School of Economics and Political Science.

Aijaz, R. (2019). *India's peri-urban regions: The need for policy and the challenges of governance*. ORF Issue Brief No. 285. Observer Research Foundation.

Aiyar, S. (2020). *The gated republic: India's public policy failures and private solutions*. Harper Collins Publishers India.

Alexander, C. (1996). *The art of being black*. Oxford University Press.

Alexander, C., Edwards, R., & Temple, B. (2007). Contesting cultural communities: Language, ethnicity and citizenship in Britain. *Journal of Ethnic and Migration Studies, 33*(5), 783–800.

Allen, A., Lampis, A., & Swilling, M. (Eds.). (2016). *Untamed urbanisms*. Routledge.

Alleyne, B. (2002). An idea of community and its discontents. *Ethnic and Racial Studies, 25*(4), 607–627.

Amarakoon, V., Trafford, J. A. P., Udeshika, T., Amarasekara, D. S., & Wickramasinghe, D. (2024). *Towards happy relief campers: Surfacing psycho-social issues, conflicts and other problems for flood-affected residents and officials in Kuruwita, Sri Lanka. International Journal of Disaster Risk Reduction, 101* [Online]. https://doi.org/10.1016/j.ijdrr.2024.104259

Amin, F., Luxmi, S., Ali, F., & Fareeduddin, M. (2023). Flood 2022 in Pakistan: Managing medical flood relief camps in a developing country. *Journal of Family Medicine and Primary Care, 12*(2), 194–200. https://doi.org/10.4103/jfmpc.jfmpc_1919_22

Anderson, K. B. (1995). *Lenin, Hegel and western Marxism: A critical study*. University of Illinois Press.

Anderson, K. B. (2020). *Dialectics of revolution*. Daraja Press.

Arendt, H. (1969). *On violence*. Harcourt Brace Jovanovich Publishers.

Arora, M., & Silva, M. (2022). *Assam: Muslims falsely accused of waging 'flood jihad*. BBC, August 3. Retrieved May 6, 2024, from https://www.bbc.com/news/world-asia-india-62378520

Ascher, K. (1987). *The politics of privatisation: Contracting out public services*. Macmillan.

Aslam, M. (2007). *Panchayati raj in India*. National Book Trust.

Baker, I., Peterson, A., Brown, G., & McAlpine, C. (2012). *Local government response to the impacts of climate change: An evaluation of local climate adaptation plans. Landscape and Urban Planning, 107*(2), 127–136.

166 References

Barman, R. P. (2023). *River, society and culture: Environmental perspectives on the rivers of Assam and Bengal*. Primus Books.

Barron, J., Crawley, G., & Wood, T. (1991). *Councilors in crisis: The public and private worlds of local councilors*. Macmillan.

Barthakur, R. (2022). Why have floods in Assam become an annual Scourge? *The Quint*, July 05, https://www.thequint.com/opinion/why-have-floods-in-assam-become-an-annual-scourge Retrieved July 29, 2024

Baruah, S. (1999). *India against itself: Assam and the politics of nationality*. Oxford University Press.

Baruah, M. (2023). *Slow disaster: Political ecology of hazards and everyday life in the Brahmaputra Valley, Assam*. Routledge.

Baruah, S. K. (2023a). Moving mountains. *The Week*, December 3.

Bastani, A. (2018). *Fully automated luxury communism*. Verso.

Bauman, Z. (1999). *Liquid modernity*. Polity.

Bauman, Z. (2001a). *Community: Seeking safety in an insecure world*. Polity.

Bauman, Z. (2001b). *The individualized society*. Polity.

Bauman, Z. (2003). *City of fears, city of hopes*. Goldsmiths College, University of London.

Beck, U. (1992). *Risk society: Towards a new modernity*. Sage.

Beck, U., & Beck-Gernsheim, E. (2001). *Individualization*. Sage.

Belanche, D., Casaló, L. V., & Rubio, M. N. (2021). *Local place identity: A comparison between residents of rural and urban communities. Journal of Rural Studies, 82*, 242–252. https://doi.org/10.1016/j.jrurstud.2021.01.003

Benjamin, W. (2002). *The arcades project*. Harvard University Press.

Benton, T. (2018). What Karl Marx has to say about today's environmental problems. *The Conversation*, June 5. Retrieved May 6, 2024, from https://theconversation.com/what-karl-marx-has-to-say-about-todays-environmental-problems-97479

Berman, M. (1983). *All that is solid melts into air: The experience of modernity*. Verso.

Berry, C. (2022). The substitutive state? Neoliberal state interventionism across industrial, housing and private pensions policy in the UK. *Competition and Change, 26*(2), 242–265.

Besser, T. L. (2009). Changes in small town social capital and civic engagement. *Journal of Rural Studies, 25*(2), 185–193.

Bhasin, M. (2008). India's role in South Asia: Perceived hegemony or reluctant leadership. *Indian Foreign Affairs Journal, 3*(4), 72–88.

Bhattacharjee, M. (2021). Into the twilight – Imagining a future for Barak Valley under the spectre of poverty, hegemony and sedition. *Barak Bulletin*, December 10. Retrieved May 6, 2024, from https://www.barakbulletin.com/en_US/into-the-twilight-imagining-a-future-for-barak-valley-under-the-spectre-of-poverty-hegemony-and-sedition/

Bhattacharya, P. (2003). Flood problem of Assam: Its causes and remedies. In P. C. Sabhapandit (Ed.), *Flood problem of Assam: Cause and remedies*. Omsons.

Bhattacharya, P. (2022). *Field notes from a waterborne land: Bengal beyond the Bhadralok*. Harper Collins.

Bhavsar, D., Tiwari, P., & Deshpande, V. (2020). *Study of municipal governance assessment frameworks*. Center for Water and Sanitation, CEPT University.

Bisen, A. (2019). *Wasted: The messy story of sanitation in India, a manifesto for change*. Pan Macmillan.

Bitzen, W. J. W., Deschenes, O., & Sanders, M. (2019). The economic impacts of natural disasters: A review of models and empirical studies. *Review of Environmental Economics and Policy, 13*(2), 167–188.

Blackshaw, T. (2005). *Zygmunt Bauman*. Routledge.

Boardman, A. E., & Vining, A. R. (1989). Ownership and performance in competitive environments: A comparison of the performance of private, mixed, and state-owned enterprises. *Journal of Law and Economics, 32*, 1–33.

Bönker, F., Libbe, J., & Wollmann, H. (2016). Remunicipalisation revisited: Long-term trends in the provision of local public services in Germany. In H. Wollmann, I. Koprić, & G. Marcou (Eds.), *Public and social services in Europe: From public and municipal to private sector provision*. Palgrave Macmillan.

Bookchin, M. (2015). *The next revolution: Popular assemblages and the promise of direct democracy*. Verso.

Bookchin, M. (2021). *From urbanization to cities: The politics of democratic municipalism*. AK Press.

Bora, A. K. (2003). Floods of the Brahmaputra in Assam: A management approach. In P. C. Sabhapandit (Ed.), *Flood problem of Assam: Cause and remedies*. Omsons.

Bose, R., & Saxena, N. C. (2016). Strife in a metro: Affirming rights to admission in the city of Delhi. In *India exclusion report 2016*. Centre for Equity Studies.

Boswell, T., & Dixon, W. J. (1993). Marx's theory of rebellion: A cross-national analysis of class exploitation, economic development, and violent revolt. *American Sociological Review, 58*(5), 681–702.

Bragg, B. (2017). *Roots, radicals and rockers: How skiffle changed the world*. Faber and Faber.

Brownhill, L. (2009). Mau Mau demand reparations from Britain for Colonial Times. *Capitalism, Nature, Socialism, 20*(2), 102–105.

Brownhill, L., & Turner, T. E. (2019). Ecofeminism at the heart of ecosocialism. *Capitalism, Nature, Socialism, 30*(1), 1–10. https://doi.org/10.1080/10455752.2019.1570650

Buck, N., Gordon, I. R., Hall, P., Harloe, M., & Kleinman, M. (2002). *Working capital: Life and labour in contemporary London*. Routledge.

Bulkeley, H., & Broto, V. C. (2013). Government by experiment? Global cities and the governing of climate change. *Transactions of the Institute of British Geographers, 38*(3), 361–375.

Bulkeley, H., & Kern, K. (2006). Local government and the governing of climate change in Germany and the UK. *Urban Studies, 43*(12), 2237–2259.

Burns, C., & Tobin, P. (2017). Environmental degradation. In V. Cooper & D. Whyte (Eds.), *The violence of austerity*. Pluto Press.

Burns, D., Hambleton, R., & Hogget, P. (1994). *The politics of decentralisation: Revitalising local democracy*. Palgrave Macmillan.

Burns, T. R. (1977). Unequal exchange and uneven development in social life: Continuities in a structural theory of social exchange. *Acta Sociologica, 20*(3), 217–245.

Burris, V. (1986). The discovery of the new middle class. *Theory and Society, 15*(3), 317–349.

Byrne, D., & Goodall, H. (2013). Placemaking and transnationalism: Recent migrants and a national park in Sydney, Australia. *Parks, 19*(1), 63–72.

Camfield, D. (2023). *Future on fire: Capitalism and the politics of climate change*. PM Press.

Castells, M. (1977). *The urban question: A Marxist approach*. MIT Press.

Chakraborty, M., & Bhandari, L. (2014). Spatial poverty in Assam. *Mint*, October 20. Retrieved May 6, 2024, from https://www.livemint.com/Opinion/r0wWgeyhHBW-fdKIfciWmTP/Spatial-poverty-in-Assam.html

Chakravarti, U. (1993). Conceptualising Brahmanical patriarchy in early India: Gender, caste, class and state. *Economic and Political Weekly, 28*(14), 579–585.

Chaturvedi, S. (2020). Pandemic exposes weaknesses in India's disaster management response. The Hindu Centre for Politics and Public Policy, September 22. Retrieved May 6, 2024, from https://www.thehinducentre.com/the-arena/current-issues/article32502100.ece

Choudhury, S. (2021). *The braided river: A journey along the Brahmaputra*. Harper Collins India.

Choudhury, S. (2023). *Northeast India: A political history*. Hurst.

Clark, T. N., & Lipset, S. M. (1991). Are social classes dying? *International Sociology, 6*, 397–410.

168   References

Cohen, I. J. (2016). *Solitary action: Acting on our own in everyday life.* Oxford University Press.

Cohn, G. (1910). Municipal socialism. *The Economic Journal, 20*(80), 561–568.

Cole, D. H. (1993). Marxism and the failure of environmental protection in eastern Europe and the U.S.S.R. *Articles by Maurer Faculty, 2101.* Retrieved May 6, 2024, from https://www.repository.law.indiana.edu/facpub/2101

Cook, M. (2019). *A river with a city problem: A history of Brisbane floods.* University of Queensland Press.

Craib, I. (2010). Fear, death and sociology. *Mortality, 8*(3), 285–295.

Cresswell, T. (2004). *Place: A short introduction.* Blackwell.

D'Souza, R. (2016). *Drowned and Damned: Colonialism and Flood Control in Eastern India.* Oxford: Oxford University Press.

D'Souza, V.S. (1970/1990). Slums in a Planned City: Chandigarh. In A. R. Desai and S. D. Pillai (Eds.), *Slums and Urbanization* (pp. 225–230). Bombay: Popular Prakashan.

Dahiya, B. (2003). Peri-urban environments and community driven development: Chennai, India. *Cities, 20*(5), 341–352.

Dardot, P., & Laval, C. (2013). *The new way of the world: On neoliberal society.* Verso.

Das, B. K., & Mitra, A. K. (2003). Flood management in Assam. In P. C. Sabhapandit (Ed.), *Flood problem of Assam: Cause and remedies.* Omsons.

Das, D. (2019). *Flood in Assam: Vulnerability and adaptation.* Mahi.

Das, G. (2002/2012). *The elephant paradigm: India wrestles with change.* Penguin.

Davis, M. (1990). *City of quartz: Excavating the future in Los Angeles.* Haymarket.

Davis, M. (1998). *Ecology of fear: Los Angeles and the imagination of disaster.* Verso.

Davis, M. (1999). A world's end: Drought, famine and imperialism (1896-1902). *Capitalism, Nature, Socialism, 10*(2), 3–46.

Davis, M. (2006). *Planet of slums.* Verso.

Davis, M. (2010). Who will build the ark? *New Left Review, 1*(61) [Online], https://newleftreview.org/issues/ii61/articles/mike-davis-who-will-build-the-ark.

Dawson, A. (2017). *Extreme cities: The peril and promise of urban life in the age of climate change.* Verso.

Dean, J. (2018). *Crowds and party.* Verso.

Deb, D. (2019). *Citizenship or humanitarian crisis: A study of the process of NRC updation in southern Assam.* MPhil Dissertation, Jadavpur University. Retrieved May 6, 2024, from http://20.198.91.3:8080/jspui/bitstream/123456789/2713/1/M.Phil%20%28International%20Relation%29%20Debasreeta%20Deb.pdf

Debord, G. (1968). *Society of the spectacle.* Pluto.

Deb Roy, S. (2021). *Social media and capitalism: People, communities and commodities.* Daraja Press.

Deb Roy, S. (2024a). *Pandemic fissures: Covid-19 and the obsolescence of freedom in India.* Routledge.

Deb Roy, S. (2024b). *The rise of the information technology society in India: Capitalism and the construction of a vulnerable workforce.* Palgrave Macmillan.

Deka, K. (2022). Why Assam is sceptical about Amit Shah's promise to make the state floods-free in five years. *India Today*, October 10. Retrieved May 6, 2024, from https://www.indiatoday.in/india-today-insight/story/why-assam-is-sceptical-about-amit-shahs-promise-to-make-the-state-floods-free-in-five-years-2283536-2022-10-10

Desai, A. R. (1969). Sociological analysis of India. In A. R. Desai (Ed.), *Rural sociology in India* (pp. 105–116). Popular Prakashan.

Desai, A. R. (1975). The myth of the welfare State. In *State and society in India: Essays in dissent.* Popular Prakashan.

Desai, A. R. (1984). *India's path of development: A Marxist approach.* Sangam Books.

Desai, A. R. (1990). Caste violence in post-partition Indian union. In A. R. Desai (Ed.), *Repression and resistance in India: Violation of the democratic rights of the working class, rural poor, Adivasis and Dalits* (pp. 363–370). Popular Prakashan.

Desai, A. R. (1991). Housing: Chaos? – Blame the victims. In A. R. Desai (Ed.), *Expanding governmental lawlessness and organised struggles: Violation of the democratic rights of the minorities, women, slum dwellers, press and some other violations.* (pp. 173–174). Popular Prakashan.

Desai, A. R., & Pillai, S. D. (1970/1990). Portrait of a Bombay slum. In A. R. Desai & S. D. Pillai (Eds.), *Slums and urbanization* (pp. 146–150). Popular Prakashan.

Desai, M. (2015). Foreword. In L. Lobo & J. Shah (Eds.), *The trajectory of India's middle class: Economy, ethics and etiquette.* Cambridge Scholars Publishing.

Deshpande, P. M. (2018). *Urban floods, experience of Navi Mumbai: Land reclamation and storm water disposal for Navi Mumbai.* Notion Press.

Dhar, S. (2015). Gender discrimination in community participation for slum development programmes: A case study of slum women in Silchar town. *International Research Journal of Social Sciences, 4*(7), 86–89.

Diaz-Quiñones, A. (2019). Foreword. In Y. Bonilla & M. LeBrón (Eds.), *Aftershocks of disaster: Puerto Rico before and after the storm.* Haymarket Books.

District Disaster Management Authority Cachar. (2022a). *Cachar flood management 2022 – An anecdote.* DMAC. Retrieved May 6, 2024, from https://cachar.gov.in/sites/default/files/swf_utility_folder/departments/cachar_epr_amtron_in_oid_2/menu/departments/flood_management_2022.pdf

District Disaster Management Authority Cachar. (2022b). *District disaster management plan: Cachar.* DDMAC. https://asdma.assam.gov.in/sites/default/files/swf_utility_folder/departments/asdma_revenue_uneecopscloud_com_oid_70/menu/document/dm_plan_cachar_2022.pdf

Diwakar, P. (2021). Spaces at risk: Urban politics and slum relocation in Chennai, India. In J. A. C. Remes & A. Horowitz (Eds.), *Critical disaster studies.* University of Pennsylvania Press.

Dorling, D. (2017). Austerity and mortality. In V. Cooper & D. Whyte (Eds.), *The violence of austerity.* Pluto Press.

Dube, E., Mtapuri, O., & Matunhu, J. (2018). Flooding and poverty: Two interrelated social problems impacting rural development in Tsholotsho District of Matabeleland North Province in Zimbabwe. *Jamba: Journal of Disaster Risk Studies, 10*(1) [Online]. https://doi.org/10.4102/jamba.v10i1.455

Dube, S. C. (1990). *Indian society.* National Book Trust.

Dunayevskaya, R. (1949). Letter to Grace Lee, February 1. In *The Raya Dunayevskaya collection – Marxist-humanism: A half century of its world development.* Wayne State University Archives of Labor and Urban Affairs.

Dunayevskaya, R. (1958/2015). *Marxism and freedom ... from 1776 until today.* Columbia University Press.

Dunayevskaya, R. (1973). *Philosophy and revolution.* Humanities Press.

Dupuis, A., & Thorne, D. C. (1998). Home, home ownership and the search for ontological security. *Sociological review, 46*(1), 24–47.

Durkheim, E. (1893/1933). *The division of labor in society.* Free Press.

Einstein, A. (1934/2012a). The question of disarmament. In J. Green (Ed.), *Albert Einstein: Selected writings.* Leftword.

Einstein, A. (1934/2012b). America and the disarmament conference of 1932. In J. Green (Ed.), *Albert Einstein: Selected writings.* Leftword.

Einstein, A. (1931/2012c). The world as I see it. In J. Green (Ed.), *Albert Einstein: Selected writings.* Leftword.

170   References

Ejaz, W., & Najam, A. (2023). The global south and climate coverage: From news taker to news maker. *Social Media + Society, 9*(2) [Online]. https://doi.org/10.1177/20563051231177904

Ellard, C. (2015). *Places of the heart: The psychogeography of everyday life.* Bellevue Literary Press.

Entrikin, N. (1991). *The betweenness of place: Towards a geography of modernity.* John Hopkins University Press.

Federici, S., & Richards, J. (2018). Every woman is a working woman. *Boston Review,* Interview, December 19. Retrieved May 6, 2024, from https://www.bostonreview.net/articles/every-woman-working-woman/

Fennema, M. (2004). The concept and measurement of ethnic community. *Journal of Ethnic and Migration Studies, 30*(3), 429–447.

Fernandes, W. (2008). *Search for peace with justice: Issues around conflicts in northeast India.* Northeastern Social Research Centre.

Fields, G. S. (2012). *Working hard, working poor: A global journey.* Oxford University Press.

Fine, G. A. (2021). *The hinge: Civil society, group cultures, and the power of local commitments.* The University of Chicago Press.

Fisher, M. (2009). *Capitalist realism.* Zero Books.

Forrest, R., & Kearns, A. (2001). Social cohesion, social capital and the neighbourhood. *Urban Studies, 38*(12), 2125–2143.

Foster, J. B. (1999). Marx's theory of metabolic rift: Classical foundations for environmental sociology. *American Journal of Sociology, 105*(2), 366–405.

Foster, J. B. (2000). *Marx's ecology: Materialism and nature.* Monthly Review Press.

Foster, J. B. (2009). *The ecological revolution: Making peace with the planet.* Monthly Review Press.

Fraser, D. (1987). Joseph chamberlain and the municipal ideal. *History Today,* April, pp. 33–39.

Furedi, F. (1997/2002). *The culture of fear: Risk-taking and the morality of low expectation.* Continuum.

G, A. H. (2022). *The stagnation of the Mexican economy is here to stay.* Elsevier eBooks. https://doi.org/10.1016/b978-0-12-815898-2.00011-2

Ge, K. (2019). *Rivers remember. #ChennaiRains and the shocking truth of a manmade flood.* Contxt.

Gehrke, J. P. (2016). A radical endeavor: Joseph chamberlain and the emergence of municipal socialism in Birmingham. *American Journal of Economics and Sociology, 75*(1), 23–57.

Gerber, L. G. (1995). Corporatism and state theory: A review essay for historians. *Social Science History, 19*(3), 313–332.

Ghosh, A. (1995). Capitalism, markets, market socialism and democracy. *Economic and Political Weekly, 30*(50), 3191–3194.

Ghosh, A. (2021). The nutmeg's curse: Parables for a planet in crisis. Penguin.

Ghosh, S. (2013). Agriculture and multidimensional poverty for human development: A case study of Barak Valley in Assam. *International Journal of Humanities and Social Sciences Invention, 2*(12), 48–57.

Ghosh, S. (2019). Is the endgame inevitable for the Ganges River Dolphin in Assam's Barak River? *The Wire,* May 3. Retrieved May 6, 2024, from https://thewire.in/environment/ganges-river-dolphin-barak-river-assam-endgame

Giddens, A. (1990). *The consequences of modernity.* Polity.

Giddens, A. (1991). *Modernity and self-identity.* Polity.

Giroux, H. (2014/2020). *Neoliberalism's war on higher education* (2nd ed.). Haymarket Books.

Gittlitz, A. M. (2018, July 14). The secret history of Marxist Alien hunters. *The Outline,* July 14, www.theoutline.com/post/5384/the-secret-history-of-marxist-alien-hunters

References 171

Gittlitz, A. M. (2020). *I want to believe: Posadism, UFOs and apocalypse communism*. Pluto.

Giuffre, K. (2013). *Communities and networks: Using social network analysis to rethink urban and community studies*. Polity.

Glaser, E. (2018). *Anti-politics: On the demonization of ideology, authority and the state*. Repeater.

Goffman, E. (1956). *The presentation of the self in everyday life*. University of Edinburgh Social Sciences Research Centre.

Goldthorpe, J. H. (1995). The service class revisited. In T. Butler & M. Savage (Eds.), *Social change and the middle classes*. Routledge.

Goonewardena, K. (2008). Marxism and everyday life: On Henri Lefebvre, Guy Debord, and some others. In K. Goonewardena, S. Kipfer, R. Milgrom, & C. Schmid (Eds.), *Space, difference and everyday life: Reading Henri Lefebvre* (pp. 117–133). Routledge.

Goonewardena, K. (2019). Monument. In Antipode Editorial Collective (Eds.), *Keywords in radical geography: Antipode at 50* (pp. 186–191). John Wiley & Sons. https://doi.org/10.1002/9781119558071.ch34

Gopalakrishnan, S. (2009). *Neoliberalism and Hindutva: Fascism, free markets and the restructuring of Indian capitalism*. Aakar.

Gordon, J. A., & Gordon, L. R. (2007). Reading the signs: A philosophical look at disaster. In K. J. Saltman (Ed.), *Schooling and the politics of disaster*. Routledge.

Government of Assam. (2022a). *Barak river system*. Govt. of Assam. Retrieved May 6, 2024, from https://waterresources.assam.gov.in/portlet-innerpage/barak-river-system

Government of Assam. (2022b). *Brahmaputra river system*. Govt. of Assam. Retrieved May 6, 2024, from https://waterresources.assam.gov.in/portlet-innerpage/brahmaputra-river-system

Government of Assam. (2022c). *Assam floods 2022: Flood memorandum to the Government of India*. Government of Assam.

Greenhalgh, C. (2005). Why does market capitalism fail to deliver a sustainable environment and greater equality of incomes? *Cambridge Journal of Economics*, *29*(6), 1091–1109.

Gu, X., Zhang, Q., Li, J., Liu, J., Xu, C., & Sun, P. (2020, May 1). The changing nature and projection of floods across Australia. *Journal of Hydrology*, 584, 124703. https://doi.org/10.1016/j.jhydrol.2020.124703

Guelke, A. (2012). *Politics in deeply divided societies*. Polity.

Guin, D. (2019). *Contemporary perspectives of small towns in India: A review. Habitat International 86*, 19027.

Gupta, A. P. (1996). Political economy of privatisation in India. *Economic and Political Weekly*, *31*(39), 2867–2894.

Gupta, A. P. (2009). Water availability, poverty and socio-economic crisis in the floodplains of Barak Valley, Assam, North East India. In H. Ayeb & T. Ruf (Eds.), *Eaux, pauvreté et crises sociales* (1–). IRD Éditions [Online]. https://doi.org/10.4000/books.irdeditions.4824

Gupta, A. P., & Sudharsanan, N. (2022). Large and persistent life expectancy disputes between India's social groups. *Population and Development Review*, *48*(3), 863–882. https://doi.org/10.1111/padr.12489

Guru, G. (2013). Limits of the organic intellectual: A Gramscian reading of Ambedkar. In C. Zene (Ed.), *The political philosophies of Antonio Gramsci and B. R. Ambedkar: Itineraries of Dalits and subalterns*. Routledge.

Guwahati Plus. (2021, July 17). Assam's below-average per-capital income as per economic survey is worrisome, says Akhil Gogoi. Retrieved May 6, 2024, from https://www.livemint.com/Opinion/r0wWgeyhHBWfdKIfciWmTP/Spatial-poverty-in-Assam.html

172 References

Haba, A. (1979). Public sector and economic policy of Asian developing countries. In *Role of state sector in developing countries*. People's Publishing House.

Hamouchene, H., & Sandwell, K. (2023). Just in time: The urgent need for a just transition in the Aran region. In H. Hamouchene & K. Sandwell (Eds.), *Dismantling green colonialism: Energy and climate justice in the Arab region*. Pluto Press.

Hardt, M., & Negri, A. (2000). *Empire*. Harvard University Press.

Hardt, M., & Negri, A. (2004). *Multitude*. Penguin.

Hardt, M., & Negri, A. (2009). *Commonwealth*. Harvard University Press.

Harriss-White, B. (2006). Poverty and capitalism. *Economic and Political Weekly, 3*(4), 1241–1246.

Harvey, D. (1979). Monument and myth. *Annals of the Association of American Geographers, 69*(3), 362–381.

Harvey, D. (1985). *Consciousness and the urban experience*. Basil Blackwell.

Harvey, D. (2003). *The new imperialism*. Oxford.

Harvey, D. (2005a). *A brief history of neoliberalism*. Oxford University Press.

Harvey, D. (2005b). *Spaces of neo liberalization: Towards a theory of uneven geographical development*. Department of Geography, University of Heidelberg.

Harvey, D. (2006). *Spaces of global capitalism*. Verso.

Harvey, D. (2007). Neoliberalism as creative destruction. *The Annals of the American Academy of Political and Social Science, 610*, 22–44.

Harvey, D. (2008). The right to the city. *New Left Review, 1*(53), 23–40.

Harvey, D. (2011). The future of the commons. *Radical History Review, 109*, 101–107.

Harvey, D. (2012). *Rebel cities: From the right to the city to the urban revolution*. Verso.

Heron, K. (2013). Capitalist catastrophism. *Roar, 10*. https://roarmag.org/magazine/capi-talist-catastrophism/

Heron, K., & Dean, J. (2020, December 2). Revolution or ruin. *E-flux Journal*. Retrieved May 6, 2024, from www.e-flux.com/journal/110/335242/revolution-or-ruin/

Heron, K., & Dean, J. (2022, June 26). Climate Leninism and revolutionary transition: Organization and anti-imperialism in catastrophic times. *Spectre Journal*. Retrieved May 6, 2024, from https://spectrejournal.com/climate-leninism-and-revolutionary-transition/

Hillier, D. (2018). *Facing risk: Options and challenges in ensuring that climate/disaster risk finance and insurance deliver for poor people*. OXFAM. Retrieved May 6, 2024, from https://oi-files-d8-prod.s3.eu-west-2.amazonaws.com/s3fs-public/file_attachments/bp-facing-risk-climate-disaster-insurance-160418-en.pdf

Hirani, S. A. A. (2022). *Caring for breastfeeding mothers in disaster relief camps: A call to innovation in nursing curriculum. Science Talks*, 4, 100089. https://doi.org/10.1016/j.sctalk.2022.100089

Hogan, M. J. (1986). Corporatism: A positive appraisal. *Diplomatic History, 10*(4), 363–372.

Hogan, M. J. (1990). Corporatism. *The Journal of American History, 77*(1), 153–160.

Holloway, J. (1995). The Abyss opens: The rise and fall of Keynesianism. In W. Bonefeld & J. Holloway (Eds.), *Global capital, national state and the politics of money*. Macmillan.

Holm, A. (2006). Urban renewal and the end of social housing: The roll out of neoliberalism in east Berlin's Prenzlauer Berg. *Social Justice, 33*(3 (105)), 114–128.

Holthaus, E. (2020). *The future earth: A radical vision for what's possible in the age of warming*. Harper One.

Hudis, P. (2021). Raya Dunayevskaya's Marxist humanism and the alternative to capitalism. *Jacobin*, June 16. Retrieved April 2, 2024, from https://jacobin.com/2021/06/raya-dunayevskaya-marxist-humanism-anti-racism-capitalism-alienation

Hunter, A. (1974). *Symbolic communities: The persistence and change of Chicago's local communities*. The University of Chicago Press.

References   *173*

IMF. (2016). *Small states' resilience to natural disasters and climate change – Role for the IMF*. IMF.

India News. (2022). Assam floods: 12 more dead, 31.5 lakh affected, Silchar still under water. *Hindustan Times*, June 29. Retrieved May 6, 2024, from https://www.hindustantimes.com/india-news/assam-floods-12-more-dead-31-5-lakh-affected-silchar-still-under-water-101656523143090.html

Ingold, T. (2011). *Being Alive: Essays on Movement, Knowledge and Description*. London: Routledge.

Iyer, K. (2021). *Landscapes of loss: The story of an Indian drought*. Harper Collins.

Jabareen, Y., Eizenberg, E., & Zilberman, O. (2017). Conceptualizing urban ontological security: 'Being-in-the-city' and its social and spatial dimensions. *Cities*, *68*, 1–7.

Jaffrelot, C. (2008). 'Why should we vote?': The Indian middle class and the functioning of the world's largest democracy. In C. Jaffrelot & P. van der Veer (Eds.), *Patterns of middle-class consumption in India and China*. Sage.

Jaffrelot, C. (2011). *Religion, caste and politics in India*. Columbia University Press.

Jalan, B. (2005). *The future of India: Politics, economics and governance*. Penguin Viking.

Jalan, B. (2019). *Resurgent India: Politics, economics and governance*. Harper Collins Publishers India.

Jalan, B. (2021). *India after liberalisation: An overview*. Harper Collins.

Jamil, G. (2017). *Accumulation by segregation: Muslim localities in Delhi*. Oxford University Press.

Jewett, R. L., Mah, S. M., Howell, N., & Larsen, M. M. (2021). Social cohesion and community resilience during COVID-19 and pandemics: A rapid scoping review to inform the united nations research roadmap for COVID-19 recovery. *International Journal of Health*, *51*(3), 325–336. https://doi.org/10.1177/0020731421997092

Jha, A., Bloch, R., & Lamond, J. (2012). *Cities and flooding: A guide to integrated urban flood risk management for the 21st century*. World Bank.

Jones, L. (2018). *Big one: How natural disasters have shaped us (and what we can do about them)*. Doubleday.

Jowett, F. W. (1907). *The socialist and the city*. George Allen.

Kalita, P. (2022). In Assam's flood-hit areas, homelessness an annual affair. *Times of India*, June 27. Retrieved May 6, 2024, from https://timesofindia.indiatimes.com/city/guwahati/in-assams-flood-hit-areas-homelessness-an-annual-affair/articleshow/92483071.cms

Kalka, I. (1991). The politics of the 'Community' among Gujarati Hindus in London. *Journal of Ethnic and Migration Studies*, *17*(3), 377–385.

Kalpagam, U. (1994). *Labour and Gender: Survival in Urban India*. New Delhi: SAGE.

Kamath, L., & Vijayabaskar, M. (2009). Limits and possibilities of middle-class associations as urban collective actors. *Economic and Political Weekly*, *44*(26/27), 368–376.

Karim, L. (2021). Women as objects of development: Neoliberalism in Bangladesh. In H. Verhiven & A. Lieven (Eds.), *Beyond liberal order: State, societies and markets in the global Indian Ocean*. Hurst and Company.

Kassarda, J. D., & Parnell, A. M. (1993). *Third world cities*. Sage.

Kelman, I. (2020). *Disaster by choice: How our actions turn natural hazards into catastrophes*. Oxford University Press.

Kern, L. (2020). *Feminist city*. Verso.

Kiers, T., & Sheldrake, M. (2021). A powerful and underappreciated ally in the climate crisis. *Guardian*, November 30,

Klein, N. (2007). *The shock doctrine: The rise of disaster capitalism*. Penguin.

Klein, N. (2017). *No is not enough: Resisting Trump's shock politics and winning the world we need*. Haymarket Books.

174 References

Klein, N. (2018). *The battle for paradise: Puerto Rico takes on the disaster capitalists.* Haymarket Books.

Knowles, C. (2003). *Race and social analysis.* Sage.

Kolbert, E. (2006). *Field notes from a catastrophe: Man, nature and climate change.* Bloomsbury.

Kosik, K. (1976). *Dialectics of the concrete: A study of man and world.* Springer.

Krichen, M., Abdalzaher, M. S., Elwekeil, M., & Fouda, M. M. (2024). *Managing natural disasters: An analysis of technological advancements, opportunities, and challenges. Internet of Things and Cyber-physical Systems, 4,* 99–109.

Krishnamurthy, R. (2022a). Bedevilled Barak: DTE reconstructs what went wrong in Katigorah during 2022 Assam floods. *Down to Earth,* February 12. Retrieved May 6, 2024, from https://www.downtoearth.org.in/news/natural-disasters/bedevilled-barak-dte-reconstructs-what-went-wrong-in-katigorah-during-2022-assam-floods-94393

Krishnamurthy, R. (2022b). Bedevilled Barak: How a Katigorah school is building resilience after 2022 Assam floods. *Down to Earth,* February 13. Retrieved May 6, 2024, from https://www.downtoearth.org.in/news/natural-disasters/bedevilled-barak-how-a-katigorah-school-is-building-resilience-after-2022-assam-floods-94422

Krishnamurthy, R. (2022c). Bedevilled Barak: How a PHC in Katigorah rebuilt itself after the 2022 Assam floods. *Down to Earth,* February 14. Retrieved May 6, 2024, from https://www.downtoearth.org.in/news/natural-disasters/bedevilled-barak-how-a-phc-in-katigorah-rebuilt-itself-after-the-2022-assam-floods-94440

Krishnamurthy, R. (2022d). Bedevilled Barak: 2022 Assam deluge damaged houses in Katigorah, pushing locals to flood-proof their dwellings. *Down to Earth,* February 16. Retrieved May 6, 2024, from https://www.downtoearth.org.in/news/natural-disasters/bedevilled-barak-2022-assam-deluge-damaged-houses-in-katigorah-pushing-locals-to-flood-proof-their-dwellings-94492

Kumar, S. (2021). Historical review of flood control in Assam. *Journal of Language and Linguistic Studies, 17*(3), 1819–1830.

Kumar-Rao, A. (2023). *Marginlands: Indian landscapes on the Brink.* Pan Macmillan.

Kuttappan, R. (2019). *Rowing between the rooftops: The heroic fishermen of the Kerala floods.* Speaking Tiger.

Laing, R. D. (1969/1990). *The divided self: An existential study in sanity and madness.* Penguin Books.

Lal, A. (2019). *Life after the floods: Reflections on the Kerala floods.* Kalamos.

La Porta, R., & de Silanes, F. L. (1999). The benefits of privatisation: Evidence from Mexico. *Quarterly Journal of Economics, 114,* 1437–1467.

Lebowitz, M. (2020). *Between capitalism and community.* Monthly Review Press.

Lee, D., & Newby, H. (1983). *The problem of sociology.* Hutchinson.

Lefebvre, H. (1975/2020). *Hegel, Marx, Nietzsche or, the realm of shadows* (Trans. David Fernbach). Verso.

Lefebvre, H. (1982). *The sociology of Marx.* Columbia University Press.

Lefebvre, H. (1991a). *The production of space.* Basil Blackwell.

Lefebvre, H. (1991b). *Critique of everyday life: Volume 1.* Verso.

Lefebvre, H. (1996). *Writings on cities.* Blackwell.

Lefebvre, H. (2002). *Critique of everyday life: Volume 2.* Verso.

Lefebvre, H. (2003). *The urban revolution.* University of Minnesota Press.

Lefebvre, H. (2008). *Critique of everyday life: Volume 3: From modernity to modernism.* Verso.

Lefebvre, H. (2016a). *Marxist thought and the city.* University of Minnesota Press.

Lefebvre, H. (2016b). *Metaphilosophy.* Verso.

Le Grand, J., & Winter, D. (1986). The middle classes and the welfare state under conservative and labour governments. *Journal of Public Policy, 6*(4), 399–430.

References 175

Leopold, E., & McDonald, D. A. (2012). Municipal socialism then and now: Some lessons for the global south. *Third World Quarterly, 33*(10), 1837–1853.

Ley, D. (2004). Transnational spaces and everyday lives. *Transactions of the Institute of British Geographers, 29*(2), new series, 151–164.

Lochbaum, D., Lyman, E., Stranahan, S. Q., & Union of Concerned Scientists. (2015). *Fukushima: The story of a nuclear disaster*. The New Press.

Losurdo, D. (2006). *Liberalism: A counter history* (Trans. Gregory Elliott). Verso.

Lowenstein, A. (2015). *Disaster capitalism: Making a killing out of a catastrophe*. Verso.

Lustick, I. (1979). Stability in deeply divided societies: Consociationalism versus control. *World Politics, 31*(3), 325–344.

Maheu, A. (2012). Urbanization and flood vulnerability in a peri-urban neighbourhood of Dakar, Senegal: How can participatory GIS contribute to flood management? In W. L. Filho (Ed.), *Climate change and the sustainable use of water resources, climate change management*. Springer.

Makhija, A. K. (2006). Privatisation in India. *Economic and Political Weekly, 41*(20), 1947–1951.

Mallet, S. (2017). *River of life, river of death: The Ganges and India's future*. Oxford University Press.

Malm, A. (2016). *Fossil capital: The rise of steam power and the roots of global warming*. Verso.

Malm, A. (2020). *Corona, climate, chronic emergency: War communism in the twenty-first century*. Verso.

Mander, H., & Singh, N. (Eds.). (2021). *This land is mine, I am of this land: CAA-NRC and the manufacture of statelessness*. Speaking Tiger.

Marx, K. (1844/1973). Economic and philosophical manuscripts of 1844. In *Marx-Engels collected works: Volume 3*. Lawrence and Wishart.

Marx, K. (1849/1977). Wage, labour and capital. In *Marx-Engels collected works: Volume 9*. Lawrence and Wishart.

Marx, K. (1867/1976). *Capital: Volume 1*. Penguin.

Marx, K. (1973). *Grundrisse*. Penguin.

Marx, K., & Engels, F. (1848/1976). Manifesto of the communist party. In *Marx-Engels collected works: Volume 6*. Lawrence and Wishart.

Massey, D. (1994). *Space, place and gender*. University of Minnesota Press.

Matthewman, S. (2015). *Disasters, risks and revelation: Making sense of our times*. Palgrave.

Mattson, G. A. (1989). Defining a sense of place in small town America. *Journal of Planning Literature, 4*(4), 394–396.

Mauro, S. E., Faber, D., & Schlegel, C. (2023). Social struggles in post-Bolsonaro Brazil: An interview with Cassia Bechara of the Movimento dos Trabalhadroes Rurais Sem Terra (MST). *Capitalism, Nature, Socialism, 34*(2), 57–68.

Mbembe, A. (2019). *Necro-politics*. Duke University Press.

McDonald, D. A., & Ruiters, G. (2005). Introduction: From public to private (to public again?). In D. A. McDonald & G. Ruiters (Eds.), *The age of commodity: Water privatisation in South Africa*. Earthscan.

McKenzie, C. (2011). Social cohesion and the unequal society. *Open University*, February 11. Retrieved May 6, 2024, from https://learn1.open.ac.uk/mod/oublog/viewpost.php?post=44524

McNeill, W. (1976). *Plagues and peoples*. Anchor Books.

Means, R., & Smith, R. (1994). *Community care: Policy and practice*. Macmillan.

Mehta, A. (2022). 'Flood Jihad': How media outlets communalised Silchar floods. *The Wire*, July 13. Retrieved May 6, 2024, from https://thewire.in/environment/flood-jihad-media-outlets-communalise-silchar-floods#:~:text=On%20June%2026%2C%20Assam%20chief,10%20kilometres%20away%20from%20Silchar

176 References

Merrifield, A. (2014). Foreword. In B. Fraser (Ed.), *Marxism and urban culture*. Lexington Books.

Merrifield, A. (2017). Fifty years on: The right to the city. In *The right to the city: A Verso report*. Verso.

Mezzadri, A. (2021). Marx's field as our global present. In A. Mezzadri (Ed.), *Marx in the field*. Anthem Press.

Miliband, R. (1964). *Parliamentary socialism: A study in the politics of labour*. Monthly Review Press.

Mitra, S., & Parakal, P. V. (1979). An Introduction to role of state sector. In *Role of state sector in developing countries*. People's Publishing House.

Mohan, T. T. R. (2015). *Rethinc: What's broke at today's corporations and how to fix it*. Random House.

Mohanty, P. K. (2016). *Financing cities in India: Municipal reforms, fiscal accountability and urban infrastructure*. Sage.

Mondal, D. (2021). Basic service provisioning in peri-urban India: A regional perspective from Kolkata metropolis. *Indian Journal of Human Development, 15*(1), 97–116.

Monti, D. J. (1999). *The American city: A social and cultural history*. Basil Blackwell.

Monti, D. J. (2000). Why cities still matter. *Society, 38*, 19–27.

Monti, D. J. (2013). *Engaging strangers: Civil rotes, civic capitalism, and public order in Boston*. Farleigh Dickinson University Press [eBook].

Moore, J. W. (2015). *Capitalism in the web of life: Ecology and the accumulation of capital*. Verso.

Nagendra, H., & Mundoli, S. (2023). *Shades of blue: Connecting the drops in India's cities*. Penguin.

Nandi, J. (2022). Assam, Meghalaya recorded highest June rainfall in 121 years: IMD. *Hindustan Times*, July 7. Retrieved May 6, 2024, from https://www.hindustantimes.com/india-news/assam-meghalaya-recorded-highest-june-rainfall-in-121-years-imd-101657103468902.html

Narain, S. (2017). *Conflicts of interest: My journey through India's green movement*. Noida: Penguin.

Nath, A., & Ghosh, S. (2022). The influence of urbanization on the morphology of the Barak River floodplain in Cachar District, Assam. *Water Policy, 24*(12), 1876–1894.

National Disaster Management Authority. (2019). *Guidelines on minimum standards of relief*. Government of India. Retrieved May 6, 2024, from https://nidm.gov.in/PDF/pubs/NDMA/19.pdf

Nayar, B. R. (2020). Economic planning after economic liberalization: Between planning commission and think tank NITI, 1991-2015. In S. Mehrotra & S. Guichard (Eds.), *Planning in the 20th century and beyond*. Cambridge University Press.

Negri, A. (2017). *Marx and Foucault: Essays: Volume 1*. Polity.

Negri, M. (1982/1988). *Archaeology and the project. Revolution retrieved: Writings on Marx, Keynes, capitalist crisis and new social subjects (1967-83)*. Red Notes.

Nehru, J. (1946/2015). *The discovery of India*. Penguin.

Nordlinger, E. A. (1972). *Conflict regulation in deeply divided societies*. Center for International Affairs.

Ostry, J. D., Loungani, P., & Furceri, D. (2016). Neoliberalism: Oversold? *Finance & Development, 53*(2) [Online]. Retrieved May 6, 2024, from https://www.imf.org/external/pubs/ft/fandd/2016/06/ostry.htm

Outlook. (2022). What is causing unprecedented floods in Assam? Climate change or sloppy management? *Outlook*, July 5. Retrieved May 6, 2024, from https://www.outlookindia.com/national/what-is-causing-unprecedented-floods-in-assam-climate-change-or-sloppy-management-news-204211

Out of the Woods. (2018). The uses of disaster. *Commune*, October 22. Retrieved May 6, 2024, from https://communemag.com/the-uses-of-disaster/

OXFAM (2023). *5 Natural disasters that beg climate action*. Retrieved May 6, 2024, from https://www.oxfam.org/en/5-natural-disasters-beg-climate-action

Palmer, J. (1986). Municipal enterprise and popular planning. *New Left Review, 1*(159), 117–124.

Pande, A. (2020). *Making India great: The promise of a reluctant global power*. Harper Collins Publishers India.

Pande, P. (2023). Revisiting shortcomings of disaster management act in light of Delhi floods. *Newsclick*, July 23. Retrieved May 6, 2024, from https://www.newsclick.in/revisiting-shortcomings-disaster-management-act-light-delhi-floods

Panitch, L., & Leys, C. (1997). *The end of parliamentary socialism: From new left to new labour*. Verso.

Parenti, M. (2016). Environment-making in the Capitalocene: Political ecology of the state. In J. W. Moore (Ed.), *Anthropocene or Capitalocene? Nature, history and the crisis of capitalism*. PM Press.

Park, E. S., & Yoon, D. K. (2022). *The value of NGOs in disaster management and governance in South Korea and Japan. International Journal of Disaster Risk Reduction, 69*, 102739. https://doi.org/10.1016/j.ijdrr.2021.102739

Park, J. T. (2015). Climate change and capitalism. *Consilience, 14*, 189–206.

Park, R. (1967). *On social control and collective behaviour*. The University of Chicago Press.

Patnaik, P. (1984). Market question and capitalist development in India. *Economic and Political Weekly, 19*(31/33), 1251–1260.

Peck, J., Theodore, N., & Brenner, N. (2009). Neoliberal urbanism: Models, moments, mutations. *The SAIS Review of International Affairs, 29*(1), 49–66.

Phalkey, R., Ranzinger, S. R., Guha-Sapir, D., & Marx, M. (2010). System's impacts of natural disasters: A systematic literature review. *Health for the Millions, 12*, 10–25.

Phukan, S. D. (2003). On the study of flood problem in Assam. In P. C. Sabhapandit (Ed.), *Flood problem of Assam: Cause and remedies*. Omsons.

Pillai, S. D. (1970/1990). Slums and squatters. In A. R. Desai & S. D. Pillai (Eds.), *Slums and urbanization* (pp. 151–170). Popular Prakashan.

Pinnock, D. (1989). Ideology and urban plain: Blueprints of a garrison city. In W. James & M. Simon (Eds.) *The angry divide: Social and economic history of the Western Cape* (pp. 150–168). David Philip.

Plagerson, S., Alfers, L., & Chen, M. (2022). Introduction: Social contracts and informal workers in the global south. In L. Alfers, M. Chen, & S. Plagerson (Eds.), *Social contracts and informal workers in the global south*. Edward Elgar Publishing.

Posadas, J. (1968). Flying saucers, the process of matter and energy, science and social- ism. In *Socialism and human relationships with nature and the cosmos*. Scientific, Cultural and Political Editions.

Posadas, J. (1969). *The revolutionary state, its transitory role and the construction of socialism*. Scientific, Cultural and Political Editions.

Posadas, J. (1978). The crisis of capitalism, war and socialism. In *War, peace, and the function of the socialist countries*. Scientific, Cultural and Political Editions.

Prasad, P. H. (1988). Roots of uneven regional growth in India. *Economic and Political Weekly, 23*(33), 1968–1992.

Pratt, M. L. (1992). *Imperial eyes: Travel writing and transculturation*. Routledge.

Price, G. (2013). India's middle-class dilemma. *The World Today, 69*(1), 6.

Purcell, M., & Tyman, S. K. (2015). Cultivating food as a right to city. *Local Environment, 20*(10), 1132–1147.

Putnam, R. D. (2000). *Bowling alone: The collapse and revival of the American community*. Simon and Schuster.

Putnam, R. D. (2007). E Pluribus Unum: Diversity and community in the twenty-first century. *Scandinavian Political Studies, 30*(2), 137–174.

178 References

Raj, S. (2016). From Marginalisation to Stereotypes – 'North East India' in Indian Media: Evidences from Focus Group Discussions in Manipur. *Journal of North East India Studies*, 6(2), 70–79.

Rajamony, V., & Mana R. N. (2022). *What we can learn from the Dutch: Rebuilding Kerala post 2018 floods*. DC Books.

Ram, R. (2014). Jawaharlal Nehru, neo-liberalism and social democracy: Mapping the shifting trajectories of developmental state in India. *Voice of Dalit*, 7(2), 187–210.

Ramesh, M. (2019). *The climate solution: India's climate change crisis and what we can do about it*. Harper Collins.

Rani, U., & Unni, J. (2004). Unorganised and organised manufacturing in India: Potential for employment generating growth. *Economic and Political Weekly*, 39(41), 4568–4580.

Rathod, B. (2023). *Dalit academic journeys: Stories of caste, exclusion and assertion in Indian higher education*. Routledge.

Reckwitz, A. (2002). Towards a theory of social practices: A development in culturalist theorizing. *European Journal of Social Theory*, 5(2), 243–263.

Rege, S. (2020). Brahmanical nature of violence against women. In S. Arya & A. S. Rathore (Eds.), *Dalit feminist theory*. Routledge.

Rodaway, P. (1994). *Sensuous geographies: Body, sense and place*. Routledge.

Rodgers, S. (2009). Urban geography: Urban growth machine. In R. Kitchin & N. Thrift (Eds.), *International encyclopaedia of human geography*. Elsevier Online.

Rothbart, M., & John, O. P. (1985). Social categorization and behavioural episodes: A cognitive analysis of the effects of intergroup contact. *Journal of Social Issues*, 41, 81–104.

Roy, A. (2014). The NGO-ization of resistance. *Towards Freedom*, September 8. Retrieved May 6, 2024, from https://towardfreedom.org/story/archives/globalism/arundhati-roy-the-ngo-ization-of-resistance/#:~:text=NGOs%20give%20the%20impression%20that,They%20alter%20the%20public%20psyche

Roychowdhury, S. (2003). Public sector restructuring and democracy: The state, labour and trade unions in India. *The Journal of Development Studies*, 39(3), 29–50.

Rustin, M. (1986). Lessons of the London industrial strategy. *New Left Review*, 1(155), 75–84.

Saikia, A. (2019). *The unquiet river*. Oxford University Press.

Sainath, P. (1996). *Everybody loves a good drought: Stories from India's poorest districts*. Penguin.

Saito, K., & Goodfellow, M. (2023). A greener Marx? Kohei Saito on connecting communism with the climate crisis. *The Guardian*, February 28. Retrieved May 6, 2024, from https://www.theguardian.com/environment/2023/feb/28/a-greener-marx-kohei-saito-on-connecting-communism-with-the-climate-crisis#:~:text="With%20his%20growing%20interest%20in,the%20central%20contradiction%20of%20capitalism."&text=By%20synthesising%20these%20different%20approaches,a%20"degrowth%20ecological%20communist

Sarker, K. (2014). Neoliberal state, austerity, and workers' resistance in India. *Interface: A Journal for and About Social Movements*, 6(1), 416–440.

Sarma, J. N. (1966). Problems of economic development in Assam. *Economic and Political Weekly*, 1(7), 281+283–286.

Sathye, M. (2005). Privatization, performance, and efficiency: A study of Indian banks. *Vikalpa: The Indian Journal of Decision Makers*, 30(1), 7–16.

Scheidel, W. (2018). *The great leveler: Violence and the history of inequality from the stone age to the twenty-first century*. Princeton University Press.

Schmitter, P. C. (1974). Still the century of corporatism. In P. C. Schmitter & G. Lehmbruch (Eds.), *Trends toward corporatist intermediation*. Thousand Oaks: Sage.

Scott, S. (2009). *Making sense of everyday life*. Polity.

References   179

Seamon, D. (1979). *A geography of lifeworld: Movement, rest and encounter*. Croom Helm.

Seamon, D. (2000). Phenomenology, place and architecture: A review of the literature. *Environment and Ecology*. Retrieved May 6, 2024, from http://environment-ecology.com/environment-and-architecture/113-phenomenology-place-environment-and-architecture-a-review-of-the-literature.html

Seamon, D., & Mugerauer, M. (Eds). (1989). *Dwelling, place and environment: Towards a phenomenology of person and the world*. Columbia University Press.

Sen, A. (2011). *Marginal on the map: Hidden wars and hidden media in north east India*. Reuters Institute Fellowship Paper. University of Oxford.

Sen, S. (2019). *Ganga: The many pasts of a river*. Penguin.

Sen, S., & Dhawan, N. (2015). Addressing domestic violence: Changing strategies within the women's movement, Kolkata, 1980-2010. In S. K. Das (Ed.), *India: Democracy and violence* (pp. 169–212). Oxford University Press.

Sen, S. K. (1994). *Working class movements in India 1885-1975*. Oxford University Press.

Sen, S.N. (1970/1990). Slums and Bustees in Calcutta. In A. R. Desai and S. D. Pillai (Eds.), *Slums and Urbanization* (pp. 206–207). Bombay: Popular Prakashan.

Sengupta, J. (2011). Need for austerity measures. *ORF*, November 1. Retrieved May 6, 2024, from https://www.orfonline.org/research/need-for-austerity-measures

Sennett, R. (1977). *The fall of the public man*. Cambridge University Press.

Sethness, J., & Clark, J. P. (2023). The quest for revolutionary love: John P. Clark interviews Javier Sethness about queer Tolstoy. *Capitalism Nature Socialism* [Online], https://doi.org/10.1080/10455752.2023.2271992.

Sharma, D., & Gayan, A. (2014). A study of flood mitigation of Assam. *Journal of Civil Engineering and Environmental Technology*, *1*(5), 5–8.

Sharma, M. (2017). *Caste and nature: Dalits and Indian environmental politics*. Oxford University Press.

Shaw, A. (2005). Peri-urban interface of Indian cities: Growth, governance and local initiatives. *Economic and Political Weekly*, *40*(2), 129–136.

Singh, G., Hasan, F., & Kasi, S. (2016). Medical relief camps in flood disaster-affected area: Experience in Jammu and Kashmir. *International Journal of Scientific Study*, *4*(5), 60–64.

Singh, R. (2009). *Contemporary ecological crisis: A Marxist view*. Aakar.

Singh, S. (2015). Disinvestment and performance of profit and loss making central public sector enterprises of India. *Indian Journal of Research*, *4*(4), 4–7.

Singhal, A., & Mathur, R. (2021). City gas distribution: Emerging potential. In V. S. Mehta (Ed.), *The next stop: Natural gas and India's journey to a clean energy future*. Gurugram: Harper Collins Publishers India.

Small, M. L. (2009). *Unanticipated gains: Origins of network inequality in everyday life*. Oxford University Press.

Soper, K. (2020). *Post-growth living*. Verso.

Staupe-Delgado, R. (2022). *Disasters and life in anticipation of slow calamity*. Routledge.

Stein, S. (2019). *Capital city: Gentrification and the real estate state*. Verso.

Steinführer, A., Vaishar, A., & Zapletalová, J. (2016). *The small town in rural areas as an underresearched type of settlement: Editors' introduction to the special issue*. European Countryside. https://doi.org/10.1515/euco-2016-0023

Stoker, G. (1991). *The politics of local government*. Macmillan.

Stromquist, S. (2023). *Claiming the city: A global history of workers' fight for municipal socialism*. Verso.

Sultana, K. (2009). *Silchar municipality: A study of its origin and development 1882-1990*. Unpublished PhD Thesis, Assam University.

Sullivan, M. (2020). *Ghosting the news: Local journalism and the crisis of American democracy*. Columbia Global Reports.

180 References

Suttles, G. D. (1973). *The social construction of communities.* The University of Chicago Press.

Swaminathan, M. (2000). *Weakening welfare: The public distribution of food in India.* Leftword Books.

Swyngedouw, E., & Heynen, N. C. (2003). Urban political ecology, justice and the politics of scale. *Antipode, 35*(5), 898–918.

Tacoli, C. (2017). *Why small towns matter: Urbanisation, rural transformations and food security.* International Institute for Environment and Development.

Tajfel, H. (1974). Social identity and intergroup behaviour. *Social Science Behaviour, 13*(2), 65–93.

Taylor-Gooby, P. (2015). Public policy futures: A left trilemma? In J. Green, C. Hay., & P. Taylor-Gooby (Eds.), *The British growth crisis: The search for a new model.* London: Palgrave Macmillan.

Teltumbde, A. (2018). *Republic of caste: Thinking equality in the time of neoliberal Hindutva.* Navayana.

Tenhunen, S., & Saavala, M. (2012). *An introduction to changing India: Culture, politics and development.* Anthem Press.

The Economist. (2022). India's regional inequality could be politically explosive. *The Economist,* October 27. Retrieved May 6, 2024, from https://www.economist.com/asia/2022/10/27/indias-regional-inequality-could-be-politically-explosive

The Weather Channel. (2022, May 28). Heavy showers flood Guwahati this week: Assam closer to ending may with nearly 60% excess rainfall. Retrieved May 6, 2024, from https://weather.com/en-IN/india/news/news/2022-05-28-heavy-rains-flood-guwa-hati-assam-to-end-may-with-60-excess

Thorat, A., & Thorat, S. (2022). Employment and the Dalit question. *Outlook,* February 11. Retrieved May 6, 2024, from www.outlookindia.com/magazine/story/india-news-employment-and-the-dalit-question/305415

Thurnheer, K. (2014). *Life beyond survival: Social forms of coping after the tsunami in war-affected eastern Sri Lanka.* Transcript Verlag.

Tiwari, P. (2019). Dynamics of peri urban areas of Indian cities. *International Journal of Scientific and Engineering Research, 10*(4), 131–141.

Tiwari, S. (2022). The worst floods in 122 years and how climate change is making them worse. *The Quint,* July 12. Retrieved May 6, 2024, from https://www.thequint.com/climate-change/assam-the-worst-floods-in-122-years-and-how-climate-change-is-making-them-worse#read-more

Towler, W. G. (1909). *Socialism in local government.* The Macmillan Company.

Tronti, M. (1962). *Factory and society.* Operaismo in English. Retrieved May 6, 2024, from https://operaismoinenglish.wordpress.com/2013/06/13/factory-and-society/

Tuan, Y.F. (1977). *Space and Place: The Perspective of Experience.* Minneapolis: University of Minnesota Press.

Tudor, A. (2003). A (macro) sociology of fear? *The Sociological Review, 51*(2), 238–258.

UNESCAP. (2023, January 4). 2022: A year when disasters compounded and cascaded. Retrieved May 6, 2024, from https://reliefweb.int/report/world/2022-year-when-disasters-compounded-and-cascaded

UN-ISDR. (2009). *Terminology: On disaster risk reduction.* Retrieved May 6, 2024, from https://www.unisdr.org/files/7817_UNISDRTerminologyEnglish.pdf

United Nations. (2003). *The challenge of slums: Global report on human settlements 2003.* Earthscan.

University of Sydney. (2022, March 23). Floods expose social inequities, and potential mental health epidemic in its wake. Retrieved May 6, 2024, from https://www.sydney.edu.au/news-opinion/news/2022/03/23/floods-expose-social-inequities-and-potential-mental-health-epi.html

Vanaik, A. (2001). The new Indian right. *New Left Review, 1*(9) [Online]. Retrieved May 6, 2024, from https://newleftreview.org/issues/ii9/articles/achin-vanaik-the-new-indian-right

Vaneigem, R. (2012). *The revolution of everyday life*. PM Press.

Vidyarthee, K. B. (2014). Trajectories of Dalits' incorporation into the Indian neo- liberal business economy. In C. Still (Ed.), *Dalits in neoliberal India: Mobility of marginalisation*. New Delhi: Routledge.

Viju, B. (2019). *Flood and fury: Ecological devastation in the Western Ghats*. Penguin.

Vincent, A. (1993). Marx and law. *Journal of Law and Society, 20*(4), 371–397.

Visvanathan, S. (2018). Socio-logic of corruption. In S. Nundy, K. Desiraju, & S. Nagral (Eds.), *Healers or predators: Healthcare corruption in India*. Oxford: Oxford University Press.

Vlachos, E. (1995). *Socio-economic impacts and consequences of extreme floods*. US-Italy research workshop on the hydrometeorology, impacts and management of extreme floods.

Von Beyme, K. (1983). Neo-corporatism: A new nut in an old shell? *International Political Science Review, 4*(2), 173–196.

Vyas, S., Hathi, P., & Gupta, A. (2022). Social disadvantage, economic inequality, and life expectancy in nine Indian states. *PNAS, 119*(10) [Online]. https://doi.org/10.1073/pnas.2109226119

Wainwright, J. (2013). *In Marx's laboratory: Critical interpretations of the Grundrisse*. Brill.

Wall, M. (1999). Factors in rural community survival: Review of insights from thriving small towns. *Great Plains Research, 9*(1), 115–135.

Wallace, R., & Wallace, R. G. (2016). The social amplification of pandemic and other disasters. In R. G. Wallace & R. Wallace (Eds.), *Neoliberal Ebola: Modeling disease emergence from finance to forest and farm*. Springer.

Warner, M. (1983). Corporatism, participation and society. *Relations Industrielles/Industrial Relations, 38*(1), 28–44.

Watkins, C. (2005). Representations of space: Spatial practices and spaces of representation: An application of Lefebvre's spatial triad. *Culture and Organization, 11*(3), 209–220.

Webb, S. (1889/1948). Historic. In B. Shaw, S. Webb, G. Wallas, The Lord Oliview, W. Clarke, A. Besant, & H. Bland (Eds.), *Fabian essays*. The Garden City Press.

Wilikilagi, V. (2010). *The mechanics of civil society contribution to natural disaster management: A focus on the Fiji Red Cross Society*. Social Science Research Network. https://doi.org/10.2139/ssrn.1628890

Winkler, J. (2002). *Space, Sound and Time: A Choice of Articles in Soundscape Studies and Aesthetics of Environment 1990-2003*. Basel: University of Basel, http://www.iacsa.eu/jw/winkler_space-sound-time_10-09-19.pdf

World Bank. (2017). *Assam – Poverty, growth, and inequality (English)*. India State Briefs. World Bank Group. http://documents.worldbank.org/curated/en/545361504000062662/Assam-Poverty-growth-and-inequality

Yates, L. (2014). Everyday politics, social practices and movement networks: Daily life in Barcelona's social centres. *The British Journal of Sociology, 66*(2), 236–258. https://doi.org/10.1111/1468-4446.12101

Zaman, R. (2022). How incessant rain and the Barak river in spate overwhelmed Assam's Silchar town. *Scroll*, June 23. Retrieved May 6, 2024, from https://scroll.in/article/1026724/how-incessant-rain-and-the-barak-river-in-spate-overwhelmed-assams-silchar-town

Zerubavel, E. (1991). *The fine line: Making distinctions in everyday life*. Chicago: The University of Chicago Press.

Index

Aadhar card, 44
Adivasis, 115
Anganwadi schemes, 145
Anthropocene, 104
Assam, 69, 101
  context of, 141
Assam government, The, 36
Assam Power Distribution Corporation
      Limited (APDCL), 30
Assam Tribune, The, 13
Austerity
  measures, 101
  policies, 90
Autonomy, 74

Barak River, The, 3, 55
Barak Valley, 7, 11, 13, 25, 35, 37,
      50, 53, 61, 66, 69, 75–76,
      82–83, 97, 103, 142
  floods in, 19
  peculiarity of, 9
  towns of, 16
Bhartiya Janta Party (BJP), 34, 75,
      83, 114
Bhuj earthquake (2002), 125
Bhura, 95n4
Big Capital, 80
Biopolitical mechanisms, 125
Black marketeers, 19
Board for Industrial and Financial
      Reconstruction, 74
Brahmaputra Valleys, 11, 54
Buildings, 88
Built environment, 119

Capital's routines, 117–123
Capitalism, 53, 66, 71, 76, 81, 89, 111,
      117, 120, 142
Capitalist development, 2, 73, 151

Capitalist models of urbanization, 152
Capitalist society, 92
Capitalist system, 80
Catastrophe, 21
Central government, 75
Chennai floods, 46
Chief Minister (CM), 69
Cities, 102, 139, 150
City spaces, 86
Climate change, 1–2, 26, 103
Climate justice, 162
Co-operative community, 149
Commodities, 64, 122
  fetishism, 64
  form, 64
Communicative mechanisms, 125
Communitarian, 76, 91
Communitarianism, 135–139
Community, The, 42
Community/communities, 42, 49, 79,
      81, 120–121, 160–161
  capitalists, 80
  formation processes, 150
  leaders, 63
Contemporary capitalism, 17, 35, 52,
      69, 79, 102, 104, 131–132
  in India, 87
Contemporary cities, 123
Contemporary society, 101, 119
Contemporary urban spaces, 79
Contractualisation, 29, 31
Corporate capital, 128
Corporatism, 102, 107–109
Corruption, 98–102
Covid-19, 21, 89, 115, 144
Critical point, 109
CSOs, 105, 128
Culture, 121
Cyclones, 11

184   Index

Dalits, 50, 97, 115
Dalmia Industries, 74
De-politicised social characteristic, 40
Democracy, 39, 100
Democratic municipalism, 137
Desperate acts of desperate classes,
    63–72
Developing societies, 41
Development, 26
Developmental negligence, 69
Dialectical approach, 123
Digital news media, 15
Disaster capitalism, 40–41, 64, 73,
    96–97, 154
    complex, 77
    corruption, and Peri-Urban India,
        98–102
    emergence of petty-disaster
        capitalism, 88–98
    frameworks, 64, 77
    society of market, 80–88
    theory, 88
Disaster communism, 153, 162
Disaster Management Act, 44
Disaster mitigation, 83, 100
Disaster NGOs, growth of, 104–111
Disasters, 11, 21, 27, 39, 55, 63, 68, 76,
    97, 119, 153
    capital's routines and 'natural'
        disasters, 117–123
    disaster-struck economy, 70
    growth of disaster NGOs, 104–111
    in India, 123
    monumental amalgamations
        during, 42–52
    objectification of lives, 111–117
    and Right to the City, 139–150
Disastrous outcomes of repressive
        monuments
    misplaced trusts and state, 36–42
    monumental amalgamations
        during disasters, 42–52
    neoliberal weakening of state, 26–36
District Disaster Management Authority
        of Cachar (DDMAC), 6, 46,
        111, 136, 142, 144

Earthquakes, 11, 118
Ecological crisis, 162
Ecological destructiveness, 151
Ecology, 112
Economic development, 98
Ecosocialism, 83
Egalitarian reimagination, 125
Electricity, 135
    sector in Assam, 30
Ethnically diverse communities, 119
Ethno-nationalism, 12
Event, 110n3
Everyday life, 32, 37, 68, 92
Exceptionalism of capitalism in
        India, 96
Exclusion of marginalised individuals, 37
Extra-parliamentary mechanisms, 140

Facebook, 133
Fear, 12
Fear of missing out (FOMO), 152
Federal Emergency Management
        Agency (FEMA), 77
Flooding in Assam, 7
Floodplains, 10n9
Floods, 1–2, 4–5, 11, 16, 25, 39, 51, 53,
        68, 85, 118, 131, 149–151
    in Barak Valley, 79
    flood-related control measures in
        Assam, 20
    Jihad, 114
Floods in Silchar (2022), 110
Full-fledged disaster capitalism, 41

Ganges River Dolphin, 14
Global capitalism, 142
Global South, 1, 25, 52
Global warming, 2
Grand claims, 84
Gross state domestic product (GSDP), 2

Harvey, David, 92, 101
Hierarchical segregation, 116
High developmental index, 76
Human society, 26, 61, 103–104, 151, 156
Humanity, 21

Hyper-atomisation, 85
Hyper-atomised social existence, 70
Hyper-vanguardist model of Indian
        state, 49

Inclusion of local administrative
        bodies, 125
India, 97, 119
Indian Meteorological Department
        (IMD), 8
Indian National Congress, 83
Industrialisation, 79
Inequality, disruptions to life and
        sustenance of, 72–78
Informal workers, 50
Informational technology, 153
Institutions, 91, 123
Integrated Child Development
        Scheme (ICDS), 145
Intergovernmental Panel on Climate
        Change, The (IPCC), 151
International Monetary Fund (IMF), 1
Internet, 135
Islamophobia, The, 114

Just Transition, 142, 156
Justice, 40

Klein, Naomi, 41, 65, 76

Labelling process functions, 62
Lefebvre, Henri, 79, 84, 101, 129
Left-wing egalitarian political
        activism, 61
Libertarian municipalism, 137
Liquid Modernity, 140n7
Livelihoods, 67
Local administrative egalitarianism, 126
Local administrative unit, 126
Local bodies, 121
Local governance, 125, 143–144, 147, 160
Local politics, 127, 143
Localism and NGOs, 127–135

Mainstream disaster capitalism, 155
Marginalisation of local politics, 129

Marginalised communities, 35, 95
Marginalised people, 82
        of Silchar, 44
Marginalised populace, 119
Marginality, 9, 52
Market, 7, 29, 40
Marx, Karl, 13, 20, 120, 163
Material security, 141
Member of Legislative Assembly
        (MLA), 48
Member of Parliament (MP), 48
Metropolitan spaces, 129
Middle class, 33–34
Misplaced trusts and State, 36–42
Mobile connectivity systems, 135
Monumental amalgamations during
        disasters, 42–52
Monumentality, 118
Monuments, 91
Mortality rates in India, 116
Multidimensional poverty, 10
Mundane activities, 84
Municipal resources, 135
Municipal socialism, 126, 126n1, 137,
        143, 159
Municipalism, 135–139
Municipalities, 126, 147
Muslim populace, The, 114
Muslims, 97, 113, 115
        in contemporary India, 114

National Disaster Management
        Authority, The (NDMA),
        44, 58, 144
Natural disasters, 2, 4, 12, 17, 21–22,
        25–26, 29, 42, 51, 56, 68,
        72, 87, 102, 107, 111,
        117–123, 125, 127, 131,
        135–139, 151, 156
Natural law of capitalism, 80
Neoliberal capitalism, 32, 53, 81–82,
        85, 144
Neoliberal disaster capitalism, 58
Neoliberal intervention, 137
Neoliberal society, 97
Neoliberal weakening of state, 26–36

186   Index

Neoliberalism, 28–29, 32, 36, 41, 44, 48–49, 70, 75, 77, 96, 101, 104, 153
New public management techniques (NPM techniques), 104
Non-exploitative community, 149
Non-governmental organizations (NGOs), 33–34, 104, 126–135
Normal capitalism, 65
Normalisation bias, 138

Objectification of lives, 111–117
Omnipresent market
    disaster capitalism, corruption, and Peri-Urban India, 98–102
    emergence of petty-disaster capitalism, 88–98
    society of market, 80–88
Ontological security, 121, 141
Other backward class (OBC), 97
Oxford Committee for Famine Relief (OXFAM), 1

Parliamentary mechanisms, 140
PDS, 106
Peri-Urban India, 98–102
Petty-disaster capitalism, 98, 117
    emergence of, 88–98
Philanthropic approaches, 155
Political change, 83
Posadas, 108, 159
Posadism, 162
Post-floods waste, 149
Predictive model, 3
Private exchange-based economy, 66
Private healthcare, 31
Private space of individuals, 138
Productivity, 109, 120
Project Suraksha, 46
Psychological security, 85, 133
Public health, 98, 104
Public housing, 157
Public sector, 29, 35, 39, 140

Radical change, 70
Radical ecological movement, 159
Radical organizations, 111
Rashtriya Swayamsevak Sangh (RSS), 114
Ravages of relief activities
    desperate acts of desperate classes, 63–72
    disruptions to life and sustenance of inequality, 72–78
    struggles within relief camps, 55–63
Rehabilitation camps, 58
Reliance Group, 74
Relief camps, 90, 99, 139
    in flood-affected areas, 57
    in Silchar, 61
    struggles within, 55–63
Relief shelters, 58
Relief-camp dwellers, 99
Remal (cyclone), 9
'Right to City' concept, 101–102
    disasters and, 139–150
Rivers, 14, 103
Routines, 86, 120, 129

Schedules, 85, 120
Sentinel, The, 13
Shibbari dam, 115
Sick Industrial Companies Act (1985), 74
Silchar, 25
Silchar floods (2022), 13, 38, 46, 87, 105, 107
Slums, 93, 98
Small entrepreneurs, 19
Small towns, 76
Social capital, 119, 152n2
Social cohesion, 109, 140
Social conflicts, 140
Social groups, 120
Social inequalities, 159
Social justice-based governmental set-up, 111
Social justice-oriented housing policy, 57
Social media, 119, 133

Social practices, 84, 123
Social theory, 103
Social totality, 157
Social transformation, 43
Socialism, 66
Society, 21, 82, 102
   of market, 80–88
Socio-political landscape in Assam, 10
South of the Global South (SGS), 25, 66, 91, 109, 128, 159
Spectacle, 68, 101, 117
State, 42, 49
   misplaced trusts and, 36–42
State Government, 126
Sub-nationalism, 12
Sundarbans, 53n1

Tendencies, 137, 162
Tilas, 17
Total human phenomenon, 158
Towns, 2, 137

Underdevelopment, 37–38
Uneven development concept, 2, 73–74

United Nations Economic and Social Commission for Asia and the Pacific (UNESCAP), 1
United Nations International Strategy for Disaster Reduction, The (UN-ISDR), 1
Urban built environment, 2
Urban centres, 101
Urban culture, 122
Urban local bodies (ULBs), 47, 80, 126–127, 137
Urban marginality, 79
Urban planning, 93
Urban spaces, 2, 148
Urbanisation, 79, 135
   in India, 56
   processes in Barak Valley, 17

Water stocking, 62
Weather, 3, 44
WhatsApp, 133
Workers, 30, 39, 50, 113
Working class, 33–34, 83

Printed and bound by CPI Group (UK) Ltd, Croydon, CR0 4YY

18/12/2024

14614483-0003